# *Connecting Faith and Science*

# *Connecting Faith and Science*

## *Philosophical and Theological Inquiries*

*Matthew Nelson Hill &*
*Wm. Curtis Holtzen, editors*

CLAREMONT STUDIES IN
SCIENCE AND RELIGION 1

**Connecting Faith and Science**
Philosophical and Theological Inquiries

1325 N. College Ave
Claremont, CA 91711

ISBN 978-1-946230-14-0

Library of Congress Cataloging-in-Publication Data

Connecting Faith and Science: Philosophical and Theological Inquiries / Matthew Nelson Hill and Wm. Curtis Holtzen

Vii + 310 pp. 13 x 23 cm. -(Claremont Studies in Science & Religion 1)
Includes bibliographical references and indexes.
ISBN 978-1-946230-14-0
1. Religion and science

BL 240.3 .C55 2017

Cover: *Benjamin Franklin Drawing Electricity from the Sky* by Benjamin West

To my father and mother, Nelson and Karen Hill. For their endless love, support, and selflessness. Thank you for being not only my parents, but also my inspiration.

*Matthew Nelson Hill*

To Wende Holtzen who inspires me to keep evolving.
This book is dedicated to you…
even though you'll never read it.

*Wm. Curtis Holtzen*

## *Table of Contents*

## *Acknowledgements*

We are incredibly grateful to the authors of this book. Not only are they sharp and thoughtful, they are a talented group that is easy to work with. We are also indebted to Tom Oord and Jim Stump for their support and dedication to the exploration of how philosophy, theology, and science intersect with one another. We especially want to thank Claremont Press for their encouragement and belief in the project. We also want to acknowledge our universities, Spring Arbor University and Hope International University, and the role they played in granting us space and time to work on this project. Special thanks to Caleb Klukos for the hard work.

Matthew is particularly grateful for the McKenna Scholars endowment fund and the McKenna family whose generosity enabled extra room to work on this project. We finally want to thank the many people who came around to advise us during the editing process, among others: Tom Phillips of Claremont who graciously accepted this project, along with Roberto Sirvent and Daniel Speak for offering extra glances and a critical eye.

*Matthew Nelson Hill*
*Wm. Curtis Holtzen*

## *Foreword*

Two of the most influential cultural forces in the world today are science and religion. Almost everyone on the planet looks to one or both of these (either explicitly or implicitly) to understand their experience and answer the big questions: Where did we come from? Where are we going? What kind of thing are we? How should we live? Science is a much more recent cultural phenomenon than religion, at least if we understand science as practiced today. If we merely mean by science the practice of looking at the world and devising explanations for what we see, it is much older and is not clearly distinguished from religion. It is only with the modern practice of limiting science to natural explanations—as opposed to the supernatural explanations of many religions—that questions are raised about how scientific and religious explanations might be related to each other.

It was the Scientific Revolution of the seventeenth and eighteenth centuries that more definitively separated out (and thereby limited) science as a distinct practice and way of obtaining knowledge from other human pursuits. And beginning about the same time we can find people reflecting on the topics that today we recognize as "science and religion." But it was not really until about the 1960s that a widespread interest developed among professional academics in issues related to the intersection of science and religion. Since then, lots of books have been written and conferences held on the topic; and even distinct academic departments and degree programs in science and religion have been formed.

Initially, the academic discussions were primarily about methodology: how do we relate the findings of science with the commitments of religion? Most influential in this was Ian Barbour's four-fold typology, which laid out the options as he saw them: conflict, independence, dialogue, and integration. Almost everyone who writes about science and religion today feels some need to distance themselves from the conflict mode in which science and religion are competing explanations. Instead they look to variations on the other options to find a way for science and religion to peacefully coexist. The question of methodology is still a live one, but the field as an academic discipline has matured and moved into other important questions. This volume is a testimony to the diversity and creativity of those who read and write about science and religion today.

Judging from polls and book sales, a good portion of the audience for science and religion books rejects theories of modern science like Big Bang cosmology and evolution. The reasons for this are complex and should not be dismissed without genuinely listening to these people and attempting to understand why such a view of the world seems so obvious them. But that is not the purpose of this volume. Its contributors take faith seriously, but they also take science seriously and look to engage these two from that perspective. The essays here model the kind of serious treatment academics desire, and their insights and conclusions are hopeful for those of us who believe science and religion both are legitimate endeavors as we seek to understand the world and our place in it.

I've gotten to know the editors over the last several years through the Wesleyan Philosophical Society, and as they note in the introduction, the impetus for this volume was one of the annual meetings of the society that was organized on the topic. To be sure, I suffer from some disciplinary bias, but it seems to me that most of the issues in science and religion are ultimately philosophical. Of course there are professionally trained scientists and theologians who have acquired the philosophical sophistication required for contributing to the conversation. But there are no paths to that proficiency besides a long and deep engagement with the philosophical issues. In this regard, the editors have assembled an exciting collection of contributors and essays. I commend them to you.

Jim Stump
Senior Editor, BioLogos

## *Introduction*

The intersection of philosophy, theology, and science is a fitting example of the complex milieu in which contemporary Christians find themselves. The enormous questions about humankind, God, and the impact that science has on this relationship can often leave us wandering—groping for truth among what some consider to be competing ideologies. Many theologians and philosophers relish in the probing of these types of metaphysical questions, while at the same time find it overwhelming to deal with new advancements in science and philosophy. Yet, as Holmes Rolston III states in his *Science and Religion*,

> The interface between science and religion is, in a certain sense, a no-man's land. No specialized science is competent here, nor does classical theology or academic philosophy really own this territory. This is an interdisciplinary zone where inquirers come from many fields. But this is a land where we increasingly must live.[1]

It is in this context that Claremont Press has ventured exploration. This volume has important contributions from seemingly disparate philosophical concepts such as pantheism, Wesleyan philosophy, and neurophilosophy, to more practical subjects such as miracles, technology, and the evolutionary origins of forgiveness. The contributing authors have tackled this

---

[1] See Holmes Rolston II, "Interfaith Peacemakers," https://www.readthespirit.com/interfaith-peacemakers/holmes-rolston-iii/ (accessed 9/19/17).

project with insight and poise as they cogently responded to many of the toughest questions in the fields of science, philosophy, and theology. It is this care and responsibility that makes this volume an important contribution to the field.

**Major Contributors and Chapters**

Since many of these chapters were birthed out of the Wesleyan Philosophical Conference, it is fitting that the book begins with Mark Mann's chapter entitled, "Wesley and the Two Books: John Wesley, Natural Philosophy, and Christian Faith." Scholars in the Wesleyan-Methodist tradition have often pointed to the so-called Quadrilateral (scripture, tradition, reason, and experience) as a distinctly "Wesleyan" approach to engaging faith and science. This paper identifies why such an approach is neither helpful for contemporary faith and science dialogue nor true to John Wesley's own thinking about and method for engaging faith and science. Wesley was not only the chief organizer and theological voice for early Methodism, but also deeply involved in the British Enlightenment development of natural science. This is evident in his own research on and advocacy of various medical remedies in the *Primitive Phsyick* and, even more explicitly, *A Survey of the Wisdom of God in Creation; Or a Compendium of Natural Philosophy.* In this popular, multi-volume text, which underwent several editions, Wesley outlines and engages much of the current scientific knowledge and theories of his day. The *Survey* exemplifies a classic 'two book' model in which God, in Christ, is revealed in complimentary ways in the two books of Scripture and Nature. This approach allows Wesley to avoid many of the pitfalls found in contemporary thought, which tend to highlight conflict between biblical faith and scientific discovery. Therefore, as this essay demonstrates, Wesley has much to offer to contemporary Christians seeking to take seriously contemporary scientific discoveries and theories while also maintaining fidelity to traditional Christian doctrine, including especially the primacy of Scripture for Christian faith and practice.

Next, Wm. Curtis Holtzen writes a defense of Alvin Plantinga in his essay titled, "Is There Anything Natural about Methodological Naturalism? An Assessment of Plantinga's Critics." Here, Holtzen explores the fact that over the last several decades Scientific Creationists and Intelligent Design theorists have argued in the courts for equal time in the classroom. The courts, however, have ruled that science, by definition, must be methodologically natural. Nonetheless the question of what demarcates science from non-science has not gone away and debates over methodological naturalism (MN) rage on. This essay is an assessment of several arguments given in response to Alvin Plantinga's challenge of the notion that science inherently depends on MN as well as his case for what he calls "Augustinian Science." Plantinga's critique is leveled against such claims that science is methodologically natural by definition, that only Duhemian science is valid science, and that without MN all authentic science would simply stop. Those defending MN, including Michael Ruse, Robert Pennock, and Stephen Pope, argue Plantinga misunderstands the nature of science and that without MN science would not properly function or perhaps exist at all. It is Holtzen's assessment, however, that the arguments made in the defense of MN are not defeaters of Plantinga's critique and that each of the arguments given in favor of MN can be understood as simply versions of the argument that science, by definition, is methodological natural.

Rem B. Edwards joins the conversation in his essay titled, "God, Miracles, Creation, Evil, and Statistical Natural Laws." Here, he argues that actual entities come first; the statistical laws of nature are their effects, not their causes. Statistical laws are mentally abstracted from their habits and are only formal, not efficient, causes. They do not make anything happen or prevent anything from happening. They evolve or change as the habits of novel creatures evolve or change. They do not control or inform us about what any individual entity is doing, only about what masses of individuals on average are doing. Thus, there is no way that traditional divine miracles could "violate" them. God has sufficient efficient-causal power to create our world ex nihilo, as in the Big Bang, and to prevent evils, which God does not do from love, justice, and moral goodness. Seven elements of a viable theodicy are outlined in his essay, the most important of which is divine justice. It would be unjust for God to prevent harms in some cases but not in all. In a universe in which God prevents all evils and thus directly controls everything, no creatures would take any responsibility for what they do. A loving and just God voluntarily and consistently limits the exercise of divine power so we can be free, responsible, and creative.

Greg Cootsona's essay, "Science and the *Sensus Divinitatis*: The Promise and Problem of the Natural Knowledge of God," connects humankind's intuitive knowledge of God to the sciences. In the first article of its Nicene Creed, the Christian church confesses that God created the universe and thus humankind. Because

God's creating this world implies all human beings possess some natural knowledge and yearning for God, the doctrine of creation situates the natural knowledge of God—and more particularly, John Calvin's *sensus divinitatis*—in its appropriate context. Cootsona asks how then does the natural knowledge of God relate to science? And what is the significance for the church? This chapter's response has seven parts. First, he argues, it outlines the natural knowledge of God and particularly Calvin's *sensus divinitatis*. The next two sections of the essay present this natural knowledge in the Bible and in subsequent theological tradition. The fourth and fifth sections move to science: first, the concept of beauty in both theology and scientific theory, and second, the specific contributions of the Cognitive Science of Religion. The sixth section offers a theological critique of the *sensus divinitatis*, and the seventh section concludes with its Christological reconstruction.

Next, John Culp addresses the theological concept of panentheism in his essay titled, "Panentheism: Hindrance of Help?" Here, he argues that advocates of panentheism propose their understanding of the nature of God's relationship to the world as a way to respond to foundational issues in the religion and science discussion. This essay examines various Christian concerns about panentheism, describes panentheistic responses to those concerns, and suggests some conclusions about the helpfulness of panentheism for Christian responses to our contemporary intellectual and cultural context. A pervasive concern about panentheism is God's identity as Ultimate. Panentheism responds to this concern by pointing out the relational nature of transcendence as involving mutual relations. This panentheistic mutual relation balances divine transcendence and immanence and preserves the claim that God is unique while embracing the intimacy of the relationship between God and the universe. Panentheism, Culp argues, does not result in a God who is unable to work in the world. Instead God's action takes place through the world rather than from outside the world. This balance of transcendence and immanence avoids the difficulties of either isolating God from the world or identifying God with the world. For Culp, panentheism offers a way to support belief in the God who was longsuffering towards Israel and forgiving when Nineveh repented.

In his essay titled, "A Process Thought Inquiry into Importance: Religion as Constructive and Postmodern," John Becker situates the current predicament of religion within a process framework and offers viable options to revitalize religion's fringe status within the public sphere. From a modernist assessment, religion is generally deemed as a subjective movement, detached from the objective world and thereby having no inherent worth. This pervasive assessment is examined in light of Alfred North Whitehead's notion of importance. Becker's examination uncovers two related findings: On the one hand, it is argued that the modernist obsession with objectivity or value-free perspectives are driven by underlying subjective motives and, on the other hand, religion has become experientially inert in modernity insofar as it fails to captivate the sensitivities of modernity. While the former pronouncement is intriguing, the problem is religion's inability to generate interest within modernity by its refusal to adapt, that is, the problem is not with modernity, but religion's reluctance to explore new modes of engagement that generate interest within the public sphere. Becker's essay invites religions and their communities to reassess their respective traditions in view of modern sensitivities in order to reinstate religion's importance.

With her essay, "Panentheism: A Potential Bridge for Scientific and Religious Dialogue," Joyce Ann Konigsburg argues that the concept of panentheism occupies a location between the diametric poles of classical theism. This position insists both assertions are true: God is ontologically unique from the world and that God and the world are equivalent. While panentheism emphasizes God's influence or divine action in the world, which is a controversial topic during science–religion dialogue, some moderate panentheistic approaches resonate positively within the scientific community. These panentheistic models respect natural laws and cosmic processes while providing novel theological insights to questions regarding divine impassibility and transcendence. Consequently, Konigsburg argues that panentheism represents a potential bridge toward mutual understanding and respect during scientific and religious dialogue.

In a change of perspective into more theological concepts, Isaac Wiegman's essay, "Divine Forgiveness and Mercy in Evolutionary Perspective," argues that in many strands of the Christian faith, the bad news is that all are unrighteous, thus all deserve divine punishment. The good news for Christian theology is that Jesus has absorbed the punishment that humankind rightfully deserves. On this view, the requirement to punish is generated from within an economy of moral exchange, the system of payback. According to Wiegman, in this system, each transgression is a debt that demands proportional repayment. Moreover, Jesus' "payment" for human sin must also be understood from within *this* system. In this essay, Wiegman explores evolutionary explanations of payback that call into question the legitimacy of the system of payback. On these theories, payback motives are a product of evolutionary pressures for self-protection and systems of payback are designed to constrain their negative effects. This requires a shift in the way Christians understand the atonement: from outside the system of payback rather than within it. On this view, the bad news is our bondage to the system itself. The good news concerns our liberation: Jesus' death in the place of sinners sends a message of self-sacrificial love that allows sinners to see their worth, forgive themselves and others, and live free of the system's requirement of payback, following in Jesus' footsteps.

Dean G. Blevins' essay makes a more pragmatic assessment. His essay, "When Neuroscientists Speak Religiously: Navigating Neuroscientific Metaphysical Claims," argues that research findings from neuroscience now contribute to various fields within Christian ministry, providing biological underpinnings for ministerial practice, and provoking theological engagement with the field. Theologians find themselves wrestling with systems of mind, culture, and religion as neuroscientists offer popular writings to the public and adapt their biological insights into larger metaphysical explanatory systems. Blevins asks practical questions about how ministers might navigate larger philosophical frameworks that intersect both with religious insight and theological convictions? Theologians may adopt dialogical approaches from within theology proper, seeking to articulate theological explanations or critiques when engaging neuroscientific claims. However, theologians and ministers might also employ a different approach, one anchored in the history of religious studies. Blevins's essay argues that using a religious studies framework, theologians can form "middle principles" that allow for a broader engagement with neuroscience and religion.

Finally, in "A Thoughtful Technophobia," Joshua Kira seeks to delineate some of the peculiarities of Heidegger's view of technology, especially as it relates to the interconnectedness of the technological mode of *being* and human life. Heidegger viewed technology as indicative of *being* which can reveal possibilities, with the incumbent danger that technological *being* simplifies human enactment into gathering into a standing-reserve. Thus, one's relationship to the world is considered only in terms of how one can use natural resources, which obscures the ontological fact that the *being* of humanity is primordially and continuously related to the world. If Heidegger is substantially correct, then this could lead to a strongly instrumental view of technology that has perils in terms of understanding all worldly relationships, especially interpersonal ones, and in terms of the altering of discourse necessary for the Christian life and mission.

*Matthew Nelson Hill*

## *Wesley and the Two Books*

## *John Wesley, Natural Philosophy, and Christian Faith*

*Mark H. Mann*

In 2013 both the theological and scientific communities could have been enthusiastically celebrating the 250th anniversary of the publication of the first edition of John Wesley's *A Survey of the Wisdom of God in the Creation: Or a Compendium of Natural Philosophy*.[1] But this is one of the lesser read of Wesley's works—perhaps slightly more attended to than his four-volume *Concise History of England*—and this auspicious occurrence went almost entirely unnoticed. This oversight is unfortunate. The *Survey* has tremendous historical value as an encyclopedia of mid-eighteenth century developments within the emerging scientific community. Additionally, the *Survey* presents us with an excellent example of how Christians during the

---

[1] There were several prints and editions of the *Survey*. The first edition, and chief source in this paper, was published in 1763 in two volumes (Bristol: William Pine), followed by a three-volume edition in 1770 (Bristol, William Pine). In 1784 Wesley produced a five-volume edition (London, J. Paramore) that included a major reworking of the earlier. After 1763, Wesley had continued his studies of natural philosophy, and felt that the *Survey* needed significant updating to align with new (or at least new to him) discoveries and theories. I utilize both of these sources herein.

Enlightenment sought to understand scientific knowledge in light of Christian faith, all through the eyes of one of the great religious thinkers and leaders of the era. But the *Survey* is not merely of interest for historical reasons, for not only does it give us insight into Wesley's understanding of the relationship between science and faith, but also highlights what his potential contribution to contemporary discussions among Christians regarding such issues as evolutionary theory and the age of the universe. In order to defend this claim, I will first briefly address the so-called Wesleyan quadrilateral, especially noting its shortcomings in addressing such issues. Then I will look to what we might call a "*Wesleyan* Two Book Theory," especially drawing upon the *Survey*, and, finally, address the implications of this "theory" for contemporary conversations about the relationship between Christian faith and science.

**Beyond the Quadrilateral**

Like many other children of the larger Wesleyan-Methodist movement, I originally came to my thinking about Wesleyan perspectives on science and faith through being taught about the quadrilateral and thinking of Scripture, tradition, reason and experience as essentially four sources of authority that, when working properly together, bring us to truth. In this way of thinking about the quadrilateral, science is connected to reason and/or experience. Such a view is often bolstered by popular modernistic accounts of knowledge that identify science with reason while equating Scripture (or revelation in general) with individual faith. So, science, which is based upon observation and experiment, is purely reasonable, and its truth is perfectly and universally available to all rational minds in some kind of absolutistic and objective way.[1] According to this framing, faith is a purely subjective and individual matter into which one "takes a leap," sometimes even "counter to" reason.[2] Of course, one might say from within this framework, what is great about us as Wesleyans—and what sets us apart from those Calvinists and fundamentalists who choose the Bible over science whenever the two are in disagreement—is that we are willing to allow our faith to be informed by reason and experience, to have our reading of Scripture (and our appropriation of Christian tradition) shaped by reason and experience. And, to the point of our conversation today, we Wesleyans therefore tend not to have a problem embracing things like Darwinism, as do all "rational" people.

But, this notion of the quadrilateral is not the way that the quadrilateral worked for Wesley, and it is certainly not the way that Albert Outler—who famously named the quadrilateral, which he later regretted—believed that it worked.[3] It is also, I believe, not a particularly helpful way of making sense of how John Wesley understood the relationship between faith and

---

[1] This is certainly the way that science and faith are treated popularly by contemporary critics of religion, such as so-called New Atheists like Richard Dawkins and Christopher Hitchens.

[2] Kierkegaard seems to treat faith this way in *Fear and Trembling* and *Concluding Postscript* as does Tertullian when asking "What does Athens have to do with Jerusalem?"

[3] Outler believed that the quadrilateral had generally been misunderstood and misconstrued by many who had begun using the model to explain Wesley. See his "The Wesleyan Quadrilateral in John Wesley," *Wesleyan Theological Journal* 21.1 (Spring 1985): 16.

"science," or what he called "Natural Philosophy."[1] In fact, I would suggest that Wesley had a far more helpful way of thinking about these issues.

For starters, the quadrilateral—at least as it functioned in Wesley—had nothing explicitly to do with science—especially if we think of science as a source of truth in some sense—but more specifically regards how Wesley read and interpreted Scripture. That is, for Wesley, Scripture, tradition, reason and experience are not properly four sources of authority in matters of Christian faith. In fact, there is only *one* true source of revelation—God, in Christ, by the Holy Spirit made known to us in Scripture. In this sense, this is what Wesley meant when he spoke of himself as a man of one book: Scripture is, in a proper sense, our sole authority in matters of Christian faith and practice. Tradition, reason and experience are not sources of authority, but instead tools for reading Scripture correctly.[2] So, tradition is not an authority so much as it is a guide provided for us by

---

[1] An example of the misconstrual of the way that the so-called quadrilateral worked in Wesley is Warren Brown's otherwise admirable attempt to put theology into a constructive conversation through his variation on the quadrilateral, which he calls the "Resonance Model." Therein Brown speaks of the various aspects of the quadrilateral (to which he adds science, thus advocating a kind of "quintrilateral") as "sources" of "authority" that must be brought together into a kind of consensus (i.e., the place in which the five find "resonance" with another) in order to bring us to "Truth." He also misreads Wesley's view of both reason and experience. Cf. his "Resonance: A Model for Relating Science, Psychology, and Faith," *Journal of Psychology and Christianity* 23.2 (2004): 110-20.

[2] Mark Mann and Ron Benefiel, "Our Wesleyan Tradition: Wesleyan Faith and Practice and the PLNU Mission," *Didache: Faithful Teaching* 12.2 (Winter 2013): http://didache.nazarene.org/index.php/volume-12-2/878-didache-v12n2-01-our-wesleyan-tradition-plnu/file.

the Church (especially the creeds of the early Church) for "interpreting" some of the complex mysteries of faith expounded in Scripture, such as the doctrine of the Trinity and the incarnation.[1] Reason, likewise, is not a source in itself against which the claims of Scripture are to be judged, but a means for working through and clarifying what Scripture is truly saying.[2] And, similarly, experience is not a source of authority for helping us to decide whether or when Scripture is true, but more properly helps us to ascertain, clarify or confirm scriptural truths when Scripture is either silent or ambivalent or there is disagreement as to how to interpret Holy Writ.[3] It is not that Wesley disparages tradition, reason or experience as important for discerning God's truth and will in matters of Christian faith and practice, but instead to put them in their proper

---

[1] See Ted A. Campbell, "The Interpretive Role of Tradition," *Wesley and the Quadrilateral: Renewing the Conversation* (Nashville: Abingdon Press, 1997), 63-76.

[2] This stands in stark contrast to the view of reason once enunciated by an Episcopalian student of mine in a class in which we were discussing Richard Hooker and the development of the Anglican Trilateral. He stated proudly that Anglicans were not completely confined to the teaching of Scripture because "we also can use our reason." This was certainly not the way that Hooker understood reason. Cf. Rebekah L. Miles, "The Instrumental Role of Reason," *Wesley and the Quadrilateral*, 107-28.

[3] The current *Methodist Book of Discipline* (Nashville: The United Methodist Publishing House, 2012) actually makes almost exactly this point: "Wesley believed that the living core of the Christian faith was revealed in Scripture, illumined by tradition, vivified in personal experience, and confirmed by reason. Scripture [however] is primary, revealing the Word of God 'so far as it is necessary for our salvation'" (77). Cf. Randy L. Maddox, "The Enriching Role of Experience," *Wesley and the Quadrilateral: Renewing the Conversation* (Nashville: Abingdon Press, 1997), ch. 5.

place in relation to the primary source of authority for Christians—the Bible. Therefore, since the so-called quadrilateral is really about interpreting Scripture, and not about relating Scripture to other authorities, what might Wesley have to offer to our reflections on the relationship between science and faith? To answer this, we turn to the *Survey.*

**Wesley, the *Survey* and the Two Books**

To begin with, the very title of the *Survey* gives us a clue to what Wesley is up to. It is both a "Compendium of Natural Philosophy" and "A Survey of the Wisdom of God in Creation." The first title is a clear implication that Wesley is interested in giving the best account possible of the view of the natural world prevalent within the budding scientific community of his day. And this is exactly what he does. On page after page we find the "latest" findings on every conceivable topic of interest to early scientists: human anatomy and physiology, the size and systemic workings of the universe, the morphology of plants and insects, the cause of meteors, earthquakes and volcanoes. Much of what he claims seems rather quaint and silly to contemporary ears, but there is also much that is surprisingly accurate. The key is that, in almost every case, he is basically describing the contemporary state of knowledge about the natural world per the late 18$^{th}$ century. In other words, although certainly highlighting where there is anything but consensus about certain theories and explanation (such as the distance between the earth and sun, or whether Newton's theories provide an effective explanation for the motions of heavenly bodies[1]), the *Survey* fairly accurately reports on the current state of knowledge within the scientific community of the mid-to-late 1700s. Wesley simply has embraced the scientific consensus of his day on nearly every point, and where he was wrong, it's because just about everyone was wrong. Moreover, Wesley realized the progressive nature of such knowledge. For example, after publishing the first edition (1763) and reading further scientific treatises that overturned claims conveyed in it, Wesley produced a second edition (1784) that expanded upon and corrected the first.

The second clear implication of the title is that Wesley doesn't see the universe as a purely natural phenomenon, as merely neutral or secular "stuff," but as an expression of the Wisdom of God. It is for this reason that I would suggest that we speak in terms of a "Wesleyan Two Book Theory." Wesley believes that nature, or creation, is revelatory of important aspects of truth about God. In an appendix to the *Survey* he will talk about Nature as a kind of witness to certain truths, believing that we can learn about certain features of the Divine nature from our study of creation.[2] This belief

---

[1] His reservations about Newton arise other contemporaneous scientific theories (viz., those of John Hutchinson) that Wesley essentially affirms on scientific grounds. Cf. Wesley, *Survey* (1763), 2:138.

[2] For Wesley, creation even points to God for those without faith, serving as a kind of foundation for basic knowledge of God. As he states, "through our reasoning upon the works of the visible creation... we form an indirect and very complex notion of [God]." In this, he echoes Aquinas's five arguments, of which Wesley's focus is the arguments from causality and degrees of perfection. Cf. *Survey* (1763), 2:207, 217.

will lead Wesley to make what we might think of as some unusual claims, such as that the morphology of the eye of a common housefly—which allows the fly to see in all directions and thus be more likely to survive when we try to swat it—is a clear demonstration of God's providential wisdom in creation.[1] In this respect, Wesley at times can come off like a contemporary advocate of Intelligent Design. Everywhere in the particularities of the created order he sees evidence of the handiwork of God, and he can be all too quick to fall back on divine intervention to explain phenomena for which there was not yet a "naturalistic" account.[2] Wesley, as was the case with many scientists of the era—including Isaac Newton, it should be added—had no problem with God-of-the-gaps arguments. But if we are to see what Wesley is up to in the *Survey* as something akin to the apologetics of a contemporary Intelligent Design proponent who argues that what we perceive to be the irreducible complexity of an eye is somehow proof of God's direct intervention in the formation of species, we are missing the point.[3] Numerous writings like this existed in Wesley's day, perhaps most famously John Ray's *Wisdom of God Manifested in the Works of Creation* (1691),[4] which Wesley

---

[1] Wesley, *Survey* (1763), 1:232.

[2] Wesley, *Survey* (1763), 1:14-15, 92-93.

[3] See, for instance, Michael Behe, *Darwin's Black Box* (New York: Free Press, 1996) and *The Edge of Evolution* (New York: Free Press, 2007) as examples.

[4] Ray's work exerted considerable influence on the development of the scientific enterprise in the generations following his death, and contemporary scientists even refer to him as the "father of natural history" for his important work on fossils and embrace of progressive creation. See, for instance, British geneticist R. J. Berry, "John Ray: Father of Natural Historians," *Science and Christian Belief* 13.1 (2001): 25-38. Berry points out that Ray's association with natural

mentions with some reservations in the preface to the *Survey*.[1] However, as Randy Maddox has indicated, Wesley actually chose to avoid the more apologetic claims of writers like Ray, preferring instead to be more modest about the theological knowledge one could garner from the study of nature. In fact, as Maddox further points out, this was exactly why Wesley would have chosen to call this a Compendium of "Natural Philosophy" rather than "Natural Theology"—it was not intended by Wesley primarily to prove the truth of any particular theological claims, but instead, as Wesley puts it in the introduction, "to display the amazing power, wisdom and goodness of the great Creator; to warm our hearts, and to fill our mouths with wonder, love and praise!"[2] That is, the point of the *Survey* was to enrich the community of the Christian faithful in its engagement with the natural world rather than to convert nonbelievers to Christian faith or provide some proof of divine intervention as an explanation for its makeup and processes.

To make sense of the true subtleties of this point, and how it may serve as a constructive model for contemporary Christian engagement with the sciences, a few words should be said about Wesley's epistemology,

---

theology, which would reach its culmination in the apologetic work of William Paley (who, claims Berry, "plundered" Ray's work), would lead to "the stature and the relevance of natural history [to become] debased by its use as a 'proof' of God's design of creation and Ray's glory and title as 'Father of Nature History'… diminished" (28).

[1] Wesley, *Survey* (1763), 1:iv.

[2] Wesley, *Survey* (1763), 1:6. Cf. Randy Maddox, "John Wesley's Precedent for Theological Engagement with the Natural Sciences," *Wesleyan Theological Journal* 44.1 (Spring 2009): 23-54.

part of which is outlined in the appendix to the *Survey*. There we find Wesley quite adamant in his embrace of empiricism—that is, all knowledge comes to us through our senses, including revelation of the truths of God.[1] So, Wesley speaks about our acquisition of such knowledge in terms of the reception of testimony, and this is why he even makes a point to note that, "Evangelical faith must be partly founded on human testimony."[2] It was human beings who wrote down the actual words of Scripture and have passed them down to us, and we come to know the truth of the claims of Scripture through the trustworthy testimony of others.[3]

But, Wesley was not a Lockean empiricist, and his brand of empiricism at first look seems rather strange to contemporary philosophical tastes. This is because, drawing upon the Cambridge Platonists, Wesley believed that, in addition to our five physical senses, we also possess spiritual senses that give us access to divine and spiritual truths otherwise invisible to us.[4] It was on

---

[1] Wesley, *Survey* (1763), 2:447.

[2] Wesley, *Survey* (1763), 2:447.

[3] Wesley, *Survey* (1763), 2:447. This coincides with his commitments to Aristotelian logic prevalent at Oxford at the time of his studies there, which, unconvincingly I believe, Rex Matthews considers the guiding influence on Wesley's epistemology. Cf. Rex Matthews, "Religion and Reason Joined: A Study in the Theology of John Wesley" (Harvard Th.D. Thesis, 1986).

[4] The influence of the Cambridge Platonists on Wesley has been well documented. See for instance, John C. English, "John Wesley's Indebtedness to John Norris," *Church History* 60 (March 1991): 55-69. John Wesley Wright has recently argued (drawing upon Richard Heitzenrater's "John Wesley and the Oxford Methodists, 1725-35," [University Microfilms, 1972], appendix iv) that Norris was the chief influence of the Cambridge Platonists, as the author whom Wesley read more than any other while at Oxford. John Wright,

the basis of our possession of such spiritual senses that Wesley would affirm our access to immediate knowledge of revelation, such as the "confirming witness of the Holy Spirit." In the *Survey,* Wesley does not speak of the spiritual senses by name,[1] but they clearly are an important backdrop for many of his claims there. And, he certainly has plenty to say about them elsewhere,[2] and what he has to say is absolutely pivotal to making sense of Wesley's purposes in the *Survey.* Simply put, because of original sin, our spiritual senses don't work in our "natural state," and we are therefore not capable of appropriately discerning the truths of God—whether they be in Scripture or Nature:

> While a man is in a mere natural state… he has, in a spiritual sense, eyes and sees not; a thick impenetrable veil lies upon them; he has ears, but hears not; he is utterly deaf to what he is most of all concerned to hear…. He has no knowledge of God; no intercourse with him; he

---

"'Use' and 'Enjoy' in John Wesley: John Wesley's Participation within the Augustinian Tradition," *Wesley and Methodist Studies* 6 (2014): 29.

[1] It is unclear why he doesn't mention them explicitly, given the nature of the *Survey* and the fact that Wesley had worked out this idea decades earlier. However, he does seem to hint at them, although in an eschatological sense, when he states: "But what then, you will say, becomes of the mysteries of the Gospel? They are all laid up safe, out of our reach, to be *immediate objects of our knowledge*, when we come to face to face" [emphasis mine]. *Survey* (1763), 2:223.

[2] See, among others, his sermons "The Great Privilege of those that Are Born of God," *Works*: 5:24-27; "The New Birth," *Works*: 6:69-70; "The Difference between Walking by Sight, and Walking by Faith," *Works:* 8:256ff. When citing Wesley's *Works* I refer to the 1872 edition of *The Works of John Wesley* (ed. Thomas Jackson; London: Wesleyan Methodist Book Room, 1872). By convention I cite the volume and page number.

> is not at all acquainted with him. He has no true knowledge of God, either spiritual or eternal.... [1]

At this point we begin to see that, for Wesley, epistemology is really soteriology.[2] The gracious presence of God quickens the spiritual senses such that the sinner can begin to see a "measure of that light, some faint glimmer, which, sooner or later, more or less, enlightens" the soul and draws those who will experience the New Birth to conviction and repentance.[3] The spiritual senses then become fully enlivened in the New Birth, granting the believer access to a vision of the world and God's activity in it simply not available to those with quiescent spiritual senses. So, apart from the enlivening presence of the Holy Spirit through prevenient grace, those in the "natural state" of sinfulness are only able to "see" the "visible" realities that Scripture is merely an artifact of the past and Nature is nothing more than the matter and processes of the physical world. But, through the power of the Holy Spirit in the New Birth, our spiritual eyes and ears are opened and our capacity to reason and understand is restored so that we can truly see and comprehend the truths of God in Scripture as well as the handiwork of God in Nature.[4] Scripture becomes for the believer the Living Word of God through which the Holy Spirit teaches us, guides us, and directs us to growth in

---

[1] Wesley, "The New Birth," *Works* 6:70.

[2] As Yoshio Noro points out, the true aim of all inquiry and understanding for Wesley is the deepening of one's "existential relationship to the will of God revealed in Christ." Yoshio Noro, "Wesley's Theological Epistemology," *The Iliff Review* 28 (1971): 60.

[3] Wesley, "On Working out Our Own Salvation," *Works*: 5:512.

[4] Wesley, *Survey* (1763), 1: iii, v-vi; 2:195, 2:229.

holiness. Nature becomes the handiwork of God whose grandeur, majesty, and loving provision are manifest in the vastness of interstellar space or the infinitely subtle intricacies of the eye of a common housefly.[1] And, it is not just the spiritual senses that are enlivened, but also our rational capacities that are repaired. That is, reason by itself is no fount of true knowledge or understanding, except in generally mundane matters. True knowledge, for Wesley, concerns divine matters ultimately invisible to both human senses and reason unless, as he puts it in "The Case of Reason Impartially Considered," it is "assisted by the Holy Ghost."[2]

If you think that Wesley's views about the enlivening of the spiritual senses and reason seems strange, you are not alone. In a certain sense such a view relies heavily on a body/soul dualism and a faculty psychology that for numerous significant reasons we tend now to reject. However, I have in recent years actually come to see this as one of the more ingenious ideas that we find in Wesley. Let us consider for a moment the ways that we have come to talk about religious belief and practice in light of developments in the sociology of knowledge and, relatedly, analytical philosophy. So, *a la* sociologists like Peter Berger, we now talk about the social construction of reality and coming to religious faith—that is, conversion—as a kind of stepping into an "alternate" social world.[3] Or, per Wittgenstein, we might talk about coming to Christian

---

[1] Wesley, *Survey* (1763), 2:192-98; 1:232.

[2] Wesley, *Works*, VI:354.

[3] See Peter Berger and Thomas Luckman's now classic, *The Social Construction of Reality: A Treatise in the Sociology of Knowledge* (New York: Anchor Books, 1966), esp. 146.

faith as engaging in a particular "language game" that has its own internal logic and way of understanding life and the world.[1] Many recent philosophers of science, often referred to as "postmoderns" in the so-called science wars, including Thomas Kuhn[2] and Paul Feyerabend[3] have come to consider of the so-called scientific worldview in a similar way—as a kind of socially constructed world which one must step into in order to understand fully and utilize successfully. If, then, we are willing to forgive Wesley for thinking of spiritual senses and rationality as literal faculties of the soul,[4] and instead think of them as a particular sensibility or perspective on life, we can begin to see that Wesley was perhaps centuries ahead of his time. As Wesley might put it, then, our Spirit-inspired encounter with the Risen Christ gives us an entirely different way of seeing and thinking about all of life, including the natural world.

One more issue remains in need of being addressed in defense of my claim that Wesley held to a "Two Book" understanding of revelation, and thus addressing a Wesleyan view of evolution, and that is

---

[1] See, for instance, Ludwig Wittgenstein, *Philosophical Investigations* (Chichester, UK: Basil Blackwell, 1953).

[2] Kuhn expresses this mostly fully and famously in *The Structure of Scientific Revolutions* (Chicago: University of Chicago Press, 1962).

[3] This is most explicitly outlined in Paul Feyerabend, *Against Method: Outline of an Anarchist Theory of Knowledge* (London: New Left Books, 1975) and *Science in a Free Society* (London: New Left Books, 1978).

[4] See, for instance, *Survey* (1763) on the perpetual motions of celestial bodies (2:151), on the distance between the earth and sun (2:153), and the efficacy of gravity (2:183), and the existence of human reason and spirituality (2:197).

Wesley's claim that he was a "man of one book"—that is, the Bible.[1] Does this claim not directly contradict my claim that Wesley was a man of two books? First, while Wesley certainly did say this on numerous occasions, we should not miss his point. Indeed, as Outler famously pointed out, whatever he might have claimed, Wesley was most certainly a man of *many* books—that is, he read as widely as is conceivable, and was willing to embrace wisdom and truth wherever it might be found.[2] Second, Wesley was clearly not committed to a view of Scripture such that it was the only truly reliable source of knowledge about the natural world. Quite the contrary: Wesley is very much open to allowing science to speak for itself, even if it *seems* to contradict Scripture. In fact, when this is the case, as with the heliocentric cosmology of Copernicus, he believes that the problem is with an overly literal reading of Scripture that fails to account for how divine revelation occurs within particular contexts with particular worldviews. As he states in the 1784 edition:

> As for those scriptural expressions which seem to contradict the earth's motion, this general answer may be made to them all, that, the scriptures were never intended to instruct us in philosophy, or astronomy; and therefore, on those subjects, expressions are not always to be taken in the literal sense, but for the most part, as accommodated to the common apprehension of mankind. Men of sense, in all ages, when not treating of the science purposely, have used common language, and it would be absurd to

---

[1] Wesley, Preface to "Sermons on Several Occasions," *Works* 5:3.

[2] Outler, "Wesleyan Quadrilateral."

adopt any other, in addressing the majority of mankind.[1]

Furthermore, the only times in the *Survey* that he resorts to biblical texts to explain natural phenomena is when the scientific community has nothing to say about the issue.[2] So, for instance, he does suggest that Scripture has it right when explaining human origins—that is, God gathered the dust of the earth and breathed life into it.[3] But he makes this claim nondogmatically, turning to the scriptural account because science had provided no clear answer to this question, as it had with other questions (such as cosmology).

Wesley is also fairly clear about what he means when he speaks of Scripture as his "one book"—it is the one book he turns to for finding the way to Heaven. As he states in the Preface to his *Sermons on Several Occasions*:

> I want to know one thing,—the way to heaven; how to land safe on that happy shore. God himself has condescended to teach me the way. For this very end He came from heaven. He hath written it down in a book. O give me that book! At any price, give me the book of God! I have it: here is knowledge enough for me. Let me be *homo unius libri*. Here then I am, far from the busy ways of men. I sit down alone; only God is here. In His presence I open, I read His book; for this end, *to find the way to heaven*.[4]

---

[1] Wesley, *Survey* (1784), 2:139.

[2] Wesley, *Survey* (1763), 1:14-15, 2:229ff.

[3] Wesley, *Survey* (1763), 2:197, 229.

[4] Wesley, *Works* 5:3. Wesley italicizes the Latin here, but the following emphasis is mine.

So, when Wesley calls himself a "man of one book" it is clear that he is not rejecting a Two Book theory. Quite the contrary, a careful reading of his *Survey* indicates that, through the eyes of faith, both Scripture and Nature reveal divine truth, even if only Scripture explicitly reveals the way of salvation.

**Conclusion: Wesley and Evolution**

What remains to be seen is how Wesley would have thought about evolutionary theory.[1] Talking about Wesley and evolution is not necessarily an easy issue to address because Wesley died eighteen years before Darwin was even born. But he gives us some clues. First, it is difficult to believe that Wesley would be a young earth creationist and completely reject evolutionary theory. As we have noted, he is quite clear in his embrace of what we might call the scientific "consensus," and when rejecting one theory or another does so generally on the basis of the science rather than

---

[1] Some have noted that Wesley affirms a kind of proto-evolutionary theory in the *Survey* based upon comments he makes at the very end of the last chapter: "The whole progress of nature is so gradual, that the entire chasm from a plant to man, is filled up with the divers kinds of creatures, rising one above another, by so gentle ascent, that the transitions from one species to another are almost insensible." However, given what follows, it is clear that when he speaks of "transitions" Wesley isn't thinking of evolution or progressive changes of any kind, but instead gradations in the great chain of being (or what Wesley calls the "scale of being" or "degrees of perfection"). In fact, his point in this section is to highlight the fact that "there is an infinitely greater space between the Supreme Being and man, than between man and the lowest insect." Therefore, this should not be understood as having anything to do with evolutionary theory. Cf. *Survey* (1763), 200.

Scripture or theology.[1] Again, he is willing to let the science speak for itself, recognizing that Scripture is not concerned with scientific matters, per se, but with theological—especially with the way of salvation. Considering the overwhelming scientific evidence for evolution and the almost universal embrace of evolution within the worldwide scientific community, it is very difficult to imagine Wesley rejecting it altogether. Indeed, it seems far more plausible, perhaps even probable, that Wesley would apply the same logic to the case with evolution and the age of the universe as he does with the Copernican revolution. That is, scriptural accounts of the natural world that *seem* to contradict the findings of natural philosophy arise from the misreading of Scripture driven by the failure to recognize the extent to which Holy Writ is revealed in the common understanding and worldview of people to whom it was revealed.[2]

However, this does not mean that Wesley would necessarily have been a theistic evolutionist. In fact, there is much to suggest that he might have embraced some form of Intelligent Design. As noted previously, Wesley had no qualms with resorting to "God of the gaps" arguments when there was no viable naturalistic explanation for any particular natural process or phenomenon. Of course, there were a lot of people in Wesley's shoes in his era—such as Isaac Newton—

---

[1] Wesley, *Survey* (1763), 2:138

[2] This is akin to the approach of many current evangelical biblical scholars, including both progressives like Peter Enns (*Inspiration and Incarnation: Evangelicals and the Problem of the Old Testament* {Grand Rapids: Baker, 2005]), and conservatives such as John Walton (*The Lost World of Genesis One: Ancient Cosmology and the Origins Debate* [Downers Grove: IVP, 2009]).

because science was still in its early development, and there remained so much unexplained that scientific inquiry had succeeded in explaining. However, it is equally as plausible that Wesley, were he to live today, would consider contemporary Intelligent Design arguments as attempts to prove theological matters that can only be seen through the eyes of faith. I tend to think this more likely—that is, that Wesley would reject contemporary Intelligent Design arguments,[1] such as those posed by Michael Behe and William Dembski, or at least avoid considering them some kind of proof for the existence and intervention of some intelligent designer in the process of evolution. As noted previously, he explicitly avoided making such arguments in the *Survey* even when drawing up the work of scholars, such as John Ray, who were making them.

What is perfectly clear, however, is that Wesley would likely embrace a kind of intelligent design at a macro level—that is, he would see the beauty and order of nature as expressive of the wisdom and glory of God in Creation. Indeed, this seems to be the primary purpose of writing the *Survey*.[2] Of course, he would recognize that such a view would only be available to believers who by faith have been given eyes to see the wonder and glory of the Creator in all of Creation. In this respect, I believe that Wesley would have considered evolution to be a potential friend to faith depending on whether or not one first has faith. In other words,

---

[1] See, for instance, Wesley, *Survey* IV:153.

[2] As Wesley states in the preface to the *Survey*: "I trust therefore the following tract may... display the amazing power, wisdom and goodness of the great creator; to warm our hearts, and to fill our mouths with wonder, love and praise." *Survey* (1763), 1:vi.

evolution can be looked at in two ways. For the nonbeliever whose spiritual sensibilities have not been enlivened by faith, evolution is merely a Godless explanation for speciation driven by some combination of natural laws and random occurrences. But, for the Christian, whose spiritual eyes have been "opened" by the Holy Spirit to the work of God in creation, evolution wonderfully reveals the very wisdom and beauty of the Creator. This is a brand of Intelligent Design quite different than that embraced by contemporary members of the Intelligent Design community, and is quite amenable to, for instance, theistic evolutionary theory. On this basis, it seems likely that Wesley would have at the very least been open to integrating evolutionary theory with Christian faith, perhaps in a way akin to those within the theistic evolution community. Indeed, that seems to me to be the most natural position for someone who, like Wesley, affirms that God's Word is revealed in two books—the Book of Scripture and the Book of Nature—and that science, for the Christian, is a God-given tool for understanding the Wisdom of God in Creation.

## *Is There Anything Natural about Methodological Naturalism?*

## *An Assessment of Plantinga's Critics*

*Wm. Curtis Holtzen*

In recent years the dispute over what makes science distinct from other forms of inquiry has been intensified. Some argue that, while science is unique, it is a land without borders. Others argue that the fence which hedges off science from all things non-science is clear and unmistakable. Those who hold that science is clearly distinct from other forms of knowledge often construct their fence out of "methodological naturalism" (MN). As defined by Paul Kurtz, MN is "a methodological principle within the context of scientific inquiry; i.e., all hypotheses and events are to be explained and tested by reference to natural causes and events. To introduce a supernatural or transcendental cause within science is to depart from naturalistic explanations. On this ground, to invoke an intelligent designer or creator is inadmissible."[1] For others, such a definition is simply an *ad hoc* wall of demarcation, erected where no borders clearly exist.

Philosopher of religion, Alvin Plantinga, sets forth the case that any demarcation built upon notions of

[1] Paul Kurtz, "Darwin Re-Crucified: Why Are So Many Afraid of Naturalism?" *Free Inquiry* 18.2 (Spring 1998): 17, quoted in Barbara Forrest, "Methodological Naturalism and Philosophical Naturalism: Clarifying the Connection," *Philo* 3.2 (2000): 8.

MN has little to be said for it. Science without MN, according to Plantinga, "suffers from the considerable disadvantage of being at present both unpopular and heretical... it also has the considerable advantage of being correct."[1] While Plantinga is correct about the unpopularity of his attacks on MN, the correctness of his case has yet to be decided.

**Plantinga and his Critics on Methodological Naturalism**

Plantinga's strategy centers on a critique of four basic arguments in support of MN. The first argument he counters is the claim that science, by definition, is methodologically natural.[2] While arguing that the matter cannot be solved by mere definitions, Plantinga also sows some seeds of doubt concerning scientific mainstays such as natural laws and notions of demarcation. Plantinga's second argument concerns the theological notion of creation having "functional integrity." At issue is the claim made by supporters of MN that God need not intervene in the workings of

---

[1] Alvin Plantinga, "Methodological Naturalism?" *Intelligent Design Creationism and Its Critics: Philosophical, Theological, and Scientific Perspectives* (ed. Robert T. Pennock; Cambridge: MIT Press, 2001), 340.

[2] Despite my best efforts I am unable to find any place in which Plantinga defines "natural" or "supernatural." In his paper "Naturalism Defeated," Plantinga defines "naturalism" as simply the belief that there are no supernatural beings. In a note he adds, "If my project were giving an analysis of philosophical naturalism, more would have to be said (precisely what, for example, is a supernatural being?); for present purposes we can ignore the niceties." See "Naturalism Defeated," a 1994 self-published paper. http://www.calvin.edu/academic/philosophy/virtual_library/articles/plantinga_calvin/naturalism_defeated.pdf (accessed 11/29/2010).

nature because God has, basically, given this world the ability to self-create. Typically those who oppose functional integrity, including Plantinga, are met with retorts that they promote a "god of the gaps" theology. Thirdly, Plantinga challenges the notion that all science must be Duhemian science[1] or methodologically natural. He responds by proposing that Duhemian science can work alongside Augustinian science (the notion that science need not be limited to natural or empirical theories and explanations) for each has their place. Lastly, Plantinga challenges the claim that all science must be natural in method otherwise science will stop when it faces a challenge. Plantinga sees the first two arguments for MN as rather weak while the latter two are much stronger.

My experience, although limited, is that advocates of MN tend to counter Plantinga by ignoring his critique. Plantinga's "Methodological Naturalism?" first appeared in 1996 and has been published on at least three different occasions.[2] Many of his arguments

---

[1] Named after Pierre Duhem, the 19th century French physicist who argued that a demarcation between physics and metaphysics was necessary if explanations of phenomena were to become autonomous to particular metaphysical theories. In a later book Plantinga labels the science committed to MN as "Simonian" after Hebert Simon, noted for his works in economics, political science, sociology, and psychology. I am unsure why he makes the switch, but it is interesting to note that Duhem was a Catholic while Simon was an atheist. For Plantinga's arguments against Simonian science see Alvin Plantinga, *Where the Conflict Really Lies: Science, Religion, & Naturalism* (Oxford: Oxford University Press), 2011, chapter 6.

[2] Plantinga's article has been published in several forms all under the title "Methodological Naturalism?" I will primarily rely on a longer version published in two parts, 18:1 and 18:2, by *Origins &*

concerning MN are older than 1996 and can be found in various writings.[1] Even though Plantinga is a formidable philosopher, many authors continue to make arguments for MN with only brief mention or even total silence concerning Plantinga's thoughts. Why this happens to be the case, however, is not of concern. It is my assessment, however, that the arguments made in the defense of MN are not defeaters of Plantinga's critique. The question is not whether MN is true, good, or practical for the scientific endeavor, but whether Plantinga's arguments against MN have been shut down. In what follows I explain each of Plantinga's arguments against MN in some detail, followed by some counterarguments to Plantinga's claims.[2] I will conclude the chapter with an assessment of whether these critiques actually defeat Plantinga's case against MN.

**True by Definition**

The first target of Plantinga's attack is the notion that MN is true by definition. Plantinga's argument is a direct response to a claim made by philosopher of

---

*Design*. Page numbers were not supplied for these papers so I have assigned page numbers for each: 18:1 1-17 and 18:2 1-19. For access go to http://www.arn.org/docs/odesign/od181/methnat181.htm and http://www.arn.org/docs/odesign/od182/methnat182.htm.

[1] See *Warrant and Proper Function* (New York: Oxford University Press, 1993) and "Naturalism Defeated," a 1994 self-published paper. For a collection of essays in response to Plantinga's writings on this subject see *Naturalism Defeated? Essays on Plantinga's Evolutionary Argument against Naturalism* (ed. James K. Beilby; Ithaca: Cornell University Press, 2002) and *Intelligent Design Creationism and Its Critics*, esp. section V.

[2] In this chapter I omit the argument concerning "functional integrity," because this dispute belongs to the domain of theology and not philosophy of science.

science Michael Ruse. Ruse argues that even if Scientific Creationism was completely successful in making its case, it would still not be science for it would not yield scientific explanations. Further, "at most, [Scientific Creationism] could prove that science shows that there can be *no* scientific explanation of origins. The Creationists believe that the world started miraculously. But miracles lie outside of science, which by definition deals only with the natural, the repeatable, that which is governed by law."[1] Plantinga sees the claim as basically arguing that all hypotheses that make reference to God (or any a supernatural being) cannot be a part of science for God is something other than natural.[2]

Plantinga raises three concerns with Ruse's claim. First he notes that the question of scientific demarcation has been an ongoing energetic debate that has apparently failed to clearly distinguish science from other human activities. For Ruse's claim to have merit there would need to be a set of "necessary and sufficient conditions for distinguishing" science from non-science.[3] Plantinga states that there is no set of conditions and that he is scarcely alone in this conclusion.

Second, Plantinga challenges Ruse's claim that there are, by definition, three properties characteristic of any science: repeatability, natural, and governed by law. Regarding repeatability, Plantinga notes that there are many scientific claims that are not repeatable. For example, the Big Bang, regardless if true, is seen as

---

[1] Michael Ruse, *Darwinism Defeated* (Reading: Addison-Wesley, 1982), 322, quoted in Plantinga, "Methodological Naturalism?" 2:1, (italics inserted by Plantinga).

[2] Plantinga, "Methodological Naturalism?" 2:2.

[3] Plantinga, "Methodological Naturalism?" 2:2.

nothing less than a scientific theory. But, as scientists currently understand the Big Bang, it is not a repeatable event. Instead, the Big Bang was a unique event that took place over thirteen billion years ago, and yet no one is suggesting that modern cosmology is not science. Plantinga does not address the second characteristic of "natural," assumedly because this is the question of the entire paper. Regarding the claim that scientific theories, by definition, must be governed by law leads Plantinga to remind Ruse that "the very existence of natural law is controversial."[1] He notes that distinguished philoso-pher of science Bas van Fraassen has argued that there are no natural laws but only "regularities." A law, according to Plantinga, "is supposed to *explain* and *ground* a regularity."[2] Plantinga ends by asking that if van Fraassen is right and there are no universal laws, only regularities, does this mean that, by definition, there is no science?[3]

The third critique Plantinga offers in response to Ruse's claim is that it is difficult to see how a dispute over definitions could solve the dilemma concerning what is or is not science. This argument would be a fair defense if the disagreement was over a verbal question—"Is the English word 'science' properly applicable to a hypothesis that makes reference to God?" Plantinga says

---

[1] Plantinga, "Methodological Naturalism?" 2:2.

[2] Plantinga, "Methodological Naturalism?" 2:2 (Italics are Plantinga's). Plantinga ponders whether the idea of natural law is a vestige from Enlightenment deism and suggests that these alleged laws or regularities might better be understood as "quantified counterfactuals of divine freedom."

[3] Plantinga suggests that there are some laws, but not everything is governed by law, and this would, perhaps, account for earthquakes, the weather, and radioactive decay.

the real question, however, is: "Could a hypothesis that makes reference to God be part of science?" This question, he argues, cannot be answered by citing a definition.[1] Plantinga gives a lengthy response to this third characteristic, but I believe the real weight of his answer comes when he writes:

> The term "science" denotes an important human activity. It is difficult or impossible to give (informative) necessary and sufficient conditions for this activity; it is not possible to say just where science ends and something else (common-sense knowledge, metaphysics, epistemology, religion) begins. However, we can describe paradigms of science, and we can say informative things about what usually or often characterizes science.[2]

Plantinga's response harkens back to the problem of demarcation. He rightly notes that there is no hard or fixed rule that separates science from non-science. All demarcation attempts have not been purely objective, but have been made paradigmatically. Here it seems that the paradigm controls the definition and it is not the definition that controls the paradigm.

The most explicit response to Plantinga is made by his target, Michael Ruse. Regarding Plantinga's example of the Big Bang being a scientific theory that is not repeatable, Ruse says "many critics of the critics have countered, there surely has to be something wrong with this argument."[3] Unfortunately, Ruse never reveals what that is. Instead, Ruse counters with another unique and

---

[1] Plantinga, "Methodological Naturalism?" 2:3.

[2] Plantinga, "Methodological Naturalism?" 2:3.

[3] Michael Ruse, "Methodological Naturalism Under Attack," *South African Journal of Philosophy* 24.1 (2005): 47.

unrepeatable event, the demise of the dinosaurs. Ruse states that even though this event is unique "the various components involved in the extinction of the dinosaurs are such that they can be brought beneath regularity."[1] In other words, while the event itself may be unique and unrepeatable, the study of the event itself can be investigated scientifically.[2]

Regarding "law" versus "regularity," Ruse says that Plantinga is simply overstating van Fraassen's point. Ruse states,

> neither van Fraassen, nor anyone else is going to deny that there are certain sorts of regularities of some kind and that these are presupposed in the activity of science. At least, if this denial is at the heart of van Fraassen's thinking, then I can only say that the response of the average scientist will be: 'News to me!'"[3]

Regarding Plantinga's claim, Ruse cautiously—almost ambivalently—agrees that there are no hard lines of demarcation. He says that there are borderline cases in which there is no clear demarcation between science and non-science. However, "this is no argument against the very idea of methodological naturalism."[4] Ruse's point seems to be that while there will always be situations or anomalies that challenge the rule of MN, this is no reason to dispense with the rule.

Lastly, Ruse defends his statement that science, by definition, is methodologically naturalistic. Ruse says that of course he is not claiming victory by means of

---

[1] Ruse, "Methodological Naturalism Under Attack," 47.

[2] While Ruse does not explain just what he means by his statement I believe I have offered a fair explanation.

[3] Ruse, "Methodological Naturalism Under Attack," 48.

[4] Ruse, "Methodological Naturalism Under Attack," 49.

word definition; that would be too easy. Ruse says he is not offering an analytic or stipulative definition of "science." Instead, he says, "What I am trying to do is offer a lexical definition: that is to say, I am trying to characterize the use of the term 'science'."[1]

**Duhemian and Augustinian Science**

As noted above, Plantinga believes the following two arguments for MN to be stronger than the one above. Plantinga's critique of MN seems to stem from his plea for a theistic or "Augustinian" science. Augustinian science begins with, and takes for granted, what is "known" by the Christian community.[2] This would include knowledge of the world as divinely and purposefully created, original sin, and humans as the *imago Dei*. Augustinian science would not compete with, but be a complement to, Duhemian science.

Duhemian science, according to Plantinga, is built upon the ideas found in Pierre Duhem's *The Aim and Structure of Physical Theory*.[3] Duhem was a committed

---

[1] Ruse, "Methodological Naturalism Under Attack," 50.

[2] The claim that the Christian community should begin science with what they "know" is based upon Plantinga's "Reformed Epistemology." Reformed Epistemology, working off the ideas of Foundationalism, argues that knowledge of God is a "properly basic belief" and does not need to be inferred by any other truth for it to be reasonable. This approach to the knowledge of God is defended by philosophers such William Alston, Nicholas Wolterstorff, and Plantinga himself. For more information on Reformed Epistemology see *Faith and Rationality: Reason and Belief in God* (ed. Alvin Plantinga and Nicholas Wolterstorff; Notre Dame: University of Notre Dame Press, 1983).

[3] Pierre Duhem, *The Aim and Structure of Physical Theory* (trans. Philip P. Wiener; Princeton: Princeton University Press, 1954). Plantinga notes that the book was first published in 1906.

Roman Catholic and serious scientist who, after being accused of allowing his Christianity to wrongly influence his physics, replied that his Christianity did not wrongly influence his physics because his faith did not influence his science at all.[1] According to Plantinga, Duhem believed that religion carried little relevance in physical theories.[2] However, while Duhem was opposed to religious doctrines influencing science, he believed that physical theories were subordinate to metaphysics. Duhem wrote that, *"if the aim of physical theories is to explain experimental laws, theoretical physics is not an autonomous science; it is subordinate to metaphysics."*[3] Plantinga interprets Duhem to be saying that, "Physical theory depends upon metaphysics in such a way that someone who doesn't accept the metaphysics involved in a given physical theory can't accept the physical theory either. And the problem with *that* is that the disagreements that run riot in metaphysics will ingress into physics, so that the latter cannot be an activity we can all work at together, regardless of our metaphysical views."[4] So, if metaphysics ultimately controls our physics in such a way that it would not allow scientific cooperation without the acceptance of their metaphysics, what are we to do? According to Duhem, a common-sense methodological naturalism was the solution. Duhem's proposal was to get all scientists to agree to

---

[1] Plantinga, "Methodological Naturalism?" 2:10.

[2] Duhem states, "Was it not a glaring fact to us, as to any man of good sense, that the object and nature of physical theory are things foreign to religious doctrines and without any contact with them?" Duhem, *Aim and Structure of Physical Theory*, 10.

[3] Duhem, *Aim and Structure of Physical Theory*, 10. Emphasis is Duhem's.

[4] Plantinga, "Methodological Naturalism?" 2:10.

work cooperatively in a way that does not expect one scientist to have to employ or accept the metaphysics of another. Scientists would be restricted then from appealing to any metaphysical or religious belief or assumption that held only limited acceptance. Duhemian science, it is argued, can be universal for it does not employ local beliefs.

While Plantinga sees this as an appealing concept, he nonetheless believes this simple idea is a "bit deceptive." Plantinga argues that, "What is really important for commonality is not the absence, from science, of hypotheses referring to God, or of metaphysics as such, or other philosophical ideas, but rather the absence of views or assumptions that divide us. If there are certain metaphysical views we all share, then there would be no reason, from this point of view, for banning those metaphysical views from science."[1] Plantinga's point is that what is really important about science is not that we dispose of all metaphysical references but that we only employ metaphysics we all agree on. At this point it does not matter whether Plantinga has accurately interpreted Duhem's proposal. The key point, here, is that Plantinga argues that any metaphysics or theology can be used in science if there is universal acceptance of the idea.[2]

Plantinga's argument is simple: First, Duhemian science, as a public science, "would be maximally

---

[1] Plantinga, "Methodological Naturalism?" 2:10.

[2] Plantinga writes in the endnote: "It isn't clear to me whether Duhem himself proposes that physics shouldn't involve *any* metaphysics, or whether he thinks only that it shouldn't involve *divisive* metaphysics. He tends to write as if it is the former he has in mind; but his arguments support only the latter" "Methodological Naturalism?" 2:19 (italics Plantinga's).

inclusive and wholly neutral in respect to world-view differences that separate us."[1] According to Plantinga, neutrality can only be achieved if we exclude atheistic propositions as thoroughly as we exclude theistic.[2] Second, the inclusion of Duhemian science should not mean the exclusion of all other kinds of science be they atomist, Cartesian, Aristotelian, or Christian. For Plantinga there is nothing improper about calling any of this "science." These would not be Duhemian science, for they would not have universal assent, but Duhemian science should not exclude others from doing their science in light of what they know given their metaphysics. To better understand Plantinga's idea, a lengthy quote is necessary:

> According to the fuller Duhemian picture, then, we would all work together on Duhemian science; but each of the groups involved—naturalists and theists, for example, but perhaps others as well—could then go on to incorporate Duhemian science into a fuller context that includes the metaphysical or religious principles specific to that group. Call this broader science "Augustinian science." Of course the motivation for doing this will vary enormously from area to area. Physics and chemistry are overwhelmingly Duhemian (of course the same might not be true for *philosophy* of physics); here perhaps Augustinian science would be for the most part

[1] Plantinga, "Methodological Naturalism? Part 2," 13.

[2] For example, Plantinga argues that much of cognitive science, theories of common ancestry, randomness in genetic variation, and so on would have to be eliminated from the sphere of public science. Further, Duhemian science, logically, should not be spoken of so much as methodological naturalism but "methodological neutralism" or "metaphysical neutralism" (2:14).

otiose. The same goes for biological sciences; surely much that goes on there could be thought of as Duhemian science. On the other hand, there are also non-Duhemian elements in the neighborhood, such as those declarations of certainty and the claims that evolutionary biology shows that human and other forms of life must be seen as a result of chance (and hence can't be thought of as designed). In the human sciences, however, vast stretches are clearly non-Duhemian; it is in these areas that Augustinian science would be most relevant and important.[1]

Plantinga paints a picture of Duhemian and Augustinian science living and working side by side.[2] His depiction of Duhemian science is very thin indeed, leaving only room for MN in the most basic of issues. Augustinian scientists would be able to appeal to the supernatural, design, or anything else the scientist's worldview gave reason to hold as a foundational and properly basic belief. For Plantinga, there is no good reason to keep the supernatural from being a proper component in any robust scientific theory. In fact, there is good reason to include it. In Plantinga's view, Duhemian science comes up short. A strictly methodologically natural science, according to Plantinga, cannot explain such things as the origin of life, cosmic

---

[1] Plantinga, "Methodological Naturalism?" 2:14.

[2] Plantinga does not address in this article what this dual world might look like. Regarding the role of education, Plantinga, in a separate article writes: "should creationism be taught in the public schools? Should evolution? The answer is in each case the same: no, neither should be taught unconditionally; but yes, each should be taught conditionally." Alvin Plantinga, "Creation and Evolution: A Modest Proposal," *Intelligent Design Creationism and Its Critics*, 790.

fine-tuning, altruistic behavior, etc.[1] Augustinian science, therefore, will provide explanations unavailable to the advocates of MN. In short, "A Christian... has a certain freedom denied her naturalist counterpart: she can follow the evidence where it leads."[2] Sure, it will not be Duhemian science, that is methodologically natural, but is that a problem?[3]

Criticism of Plantinga's position has sometimes been theological in nature, such as Ernan McMullin's argument that it would be unlikely that God would intervene in such a way as to disrupt natural laws or miraculously create new organisms.[4] My concern, however, is philosophical in nature. Much of the critique here is similar to the discussion above concerning the proper definition of "science." Can Augustinian science rightly be called "science" if it is not only non-natural, but radically different from mainstream science or "science without qualification?"[5] Ruse argues that there needs to be a way to determine whether Augustinian science is as good as science without qualification. How do we determine which explanation is better than the other? This is important in practical matters as well, such as grant awards, university support, and the public classroom.[6]

---

[1] Plantinga, "Methodological Naturalism?" 1:2-13.

[2] Plantinga, "Methodological Naturalism?" 1:7.

[3] For more on Plantinga's call for Augustinian science see "Science: Augustinian or Duhemian," *Faith and Philosophy* 13.3 (1996): 368-94.

[4] Ernan McMullin, "Plantinga's Defense of Special Creation," *Christian Scholar's Review* 21.1 (1991): 55-79.

[5] This is how Michael Ruse refers to mainstream or regular science. Ruse, "Methodological Naturalism Under Attack," 51.

[6] Ruse, "Methodological Naturalism Under Attack," 51.

Robert Pennock argues that Plantinga's Augustinian science does little more than provide one additional battleground for disputes—this time theological rather than scientific. He argues that battles "over what may be presumed as 'True Christianity' with regard to 'theological facts' should give us sufficient reason to doubt whether revelation could possibly supply the purported unified basis for such a science."[1] His main point is that appeal to something other than the neutrality of MN is bound to result in more disputes and less scientific success. Most interesting is Pennock's argument that to "apply natural knowledge to understand supernatural powers, then, by definition, they would not be supernatural."[2] The irony here, according to Pennock's line of thinking, is that Plantinga, in using God as an explanatory theory for naturalistic phenomenon, actually naturalizes God. God or the supernatural, according to Pennock, is inherently mysterious and God's ways, therefore, are not privy to scientific exploration. If God is somehow used as a scientific explanation then God is no longer supernatural, but merely natural.

Finally, Stephen J. Pope argues that Plantinga's case for an Augustinian science based upon the incompleteness of MN plainly fails. Simply because scientists employing MN have not discovered all they have set out to discover is no reason to think that MN is a failed approach or that Augustinian science would be more successful. "Failure to find a cure for a certain kind of cancer does not make it scientifically legitimate to seek

---

[1] Robert T. Pennock, *Tower of Babel: The Evidence against the New Creationism* (Cambridge: MIT Press, 1999), 204.

[2] Pennock, *Tower of Babel*, 290.

out the ministrations of a witch doctor."[1] In short, MN has been fruitful regarding inquiry of the origin of species where Augustinian science has not.

**Science Stoppers**

Lastly, Plantinga addresses the claim that without MN science stops. The accusation is that when God is interjected into a theory as an explanatory force, the science must stop for God is beyond empirical observation. To claim that God is directly responsible for particular events or phenomena (human life, the eye, flagellum, light, etc.) means that there is nothing more science can do for further explanation. If God is the direct cause then what more is there to say? Once again, Plantinga believes this is a formidable issue but hardly a defense of MN.

Plantinga's response is simple, straightforward, and found within a single paragraph:

> The claim that God has directly created life (for example) may be a science stopper; it doesn't follow that God *didn't* directly create life. Obviously we have no guarantee that God has done everything by way of employing secondary causes, or in such a way as to encourage further scientific inquiry, or for our convenience as scientists, or for the benefit of the NSF [National Science Foundation]. Clearly we can't sensibly insist in advance that whatever we are confronted with is to be explained in terms of something *else* God did; he must have done *some* things directly. It would be very much worth knowing (if possible) which things he *did*

---

[1] Stephen J. Pope, *Human Evolution and Christian Ethics* (Cambridge, UK: Cambridge University Press, 2007), 73.

> do directly; to know this would be an important part of a serious and profound knowledge of the universe. The fact that such claims are science stoppers means that as a general rule they won't be helpful; it doesn't mean that they are never true, and it doesn't mean that they can never be part of a proper scientific theory. (And of course it doesn't even bear on the other ways in which Christianity or Christian theism can be relevant to science.) It is a giant and unwarranted step from the recognition that claims of direct divine activity are science stoppers to the insistence that science must pretend that the created universe is just there, refusing to recognize that it is indeed *created*.[1]

Plantinga's response to the criticism seems to be a "so what?" If it is in fact true that God has acted directly, then shouldn't we expect science to come to a stop at some point? While theologically we might be able to debate Plantinga's claim that God must have done something directly, philosophically, that is not the issue. Plantinga wants to raise doubts as to whether naturalistic explanations are logically the only way to explain the physical world. Must we commit ourselves, *a priori,* to the belief that every physical event or phenomenon can only be explained scientifically?

To remove MN as an *a priori* rule of thumb does not, according to Plantinga, bring science to a stop. Such *a priori* reasoning about the world, in his opinion, is armchair science, "trying to infer from first principles how many teeth there are in a horse's mouth."[2] The same should go for issues regarding God's action in the

---

[1] Plantinga, "Methodological Naturalism?" 2:16.

[2] Plantinga, "Methodological Naturalism?" 2:16.

world, and scientists should "rely less upon *a priori* theology and more upon empirical inquiry."[1] Again, it should be remembered, Plantinga is not saying every scientist should come to the conclusion that God has acted in such and such a manner. His point, as I understand it, is that making appeals to God's direct action can logically be done by those engaged in Augustinian science. Duhemian science can—and should—continue in the form of methodological naturalism, but this need not mean that Christian scientists must stop *their* science from recognizing God's action in the world. Likewise, it may simply be the case that those practicing Duhemian science will come to science stoppers on their own. Once again, is there any *a priori* rule that everything investigated by science not only should, but must have a natural explanation?

Plantinga concludes his paper by making an appeal to Christians to remember that, "human history is dominated by a battle, a contest between the *Civitas Dei* and the City of Man."[2] The Christian academic community is to pursue the intellectual life as citizens of the *Civitas Dei,* bringing all that they know, including what is known by faith, to their intellectual endeavors. Likewise, it has been shown that MN, when carefully examined, is composed of seemingly weak arguments. "We should therefore reject it, taken in its full generality. Perhaps we should join others in Duhemian science; but we should also pursue our own Augustinian science."[3]

There are plenty of statements made which basically claim that denying MN amounts to a "science

---

[1] Plantinga, "Methodological Naturalism?" 2:16.

[2] Plantinga, "Methodological Naturalism?" 2:17.

[3] Plantinga, "Methodological Naturalism?" 2:17.

stopper." For example, geologist Keith Miller argues that the conclusion, "God did it," would mean "prematurely terminating the search for cause-and-effect descriptions when none are yet known [and] any motivation for further research would end."[1] According to Pennock, appeal to the supernatural for explanation is simply "too easy." "One would always be able to call upon the gods for quick theoretical assistance in any circumstance."[2]

Ruse believes that the argument that scientific inquiries would, without MN, prematurely come to an end, is perhaps the strongest argument for MN. Unlike Pennock's statements above, Ruse is attempting to address the issue without assuming MN to be true *a priori*. Ruse, in a direct response to Plantinga's argument, states that history shows that the "methodologically naturalistic approach yielded fantastic dividends."[3] Ruse continues this line of defense stating, "Plantinga altogether underestimates the power and success of methodological naturalism. He can be so slighting of its potential only because he does not take modern science seriously."[4] He continues, saying that Plantinga's mind has been made up before he starts. After citing several passages from Plantinga's works he says, "Plantinga is naïve and ignorant concerning the fossil record,

---

[1] Keith B. Miller, "Design and Purpose within an Evolving Creation," *Darwinism Defeated? The Johnson-Lamoureux Debate on Biological Origins* (ed. Phillip E. Johnson, Denis O. Lamoureux, *et al.*; Vancouver: Regent College, 1999), 113.

[2] Pennock, *Tower of Babel*, 292.

[3] Ruse, "Methodological Naturalism Under Attack," 54. Ruse's example of methodologically naturalistic success is the explanation of "insect sociality in terms of individual genetic selfishness" (54).

[4] Ruse, "Methodological Naturalism Under Attack," 55.

Cambrian explosion, and in short, is a sophisticated philosopher of religion" but Ruse questions whether Plantinga's scholarship outside the field of religion is "competent."[1] Ruse, however, never addresses why science can never stop.

**Assessment of Arguments**

Judge John E. Jones, presiding judge in the Kitzmiller v. Dover trial, ruled "rigorous attachment to 'natural' explanations is an essential attribute to science by definition and by convention."[2] If the U.S. courts were the final say on all things philosophical, then Michael Ruse's response to Plantinga would stand and the whole matter would be settled. However, the courts do not rule philosophy.

It is difficult to see how this matter can be settled by a definition of words for a number of reasons. First of all, Larry Laudan notes that, from Plato to Popper, philosophers have attempted and failed to deliver a clear line of demarcation. "Whatever the strengths and deficiencies of numerous well-known efforts at demarcation… it is probably fair to say that there is no demarcation line between science and non-science, or between science and pseudo-science, which would win assent from a majority of philosophers. Nor is there one which *should* win acceptance from philosophers or anyone else."[3] The failed history of attempts at definition

---

[1] Ruse, "Methodological Naturalism Under Attack," 55-58.

[2] John E. Jones, Kitzmiller v. Dover court documents, page 66. http://www.pamd.uscourts.gov/kitzmiller/kitzmiller_342.pdf (accessed 11/19/2010).

[3] Larry Laudan, "The Demise of the Demarcation Problem," *Physics, Philosophy and Psychoanalysis: Essays in Honor of Adolf*

and demarcation certainly helps Plantinga's case. Secondly, Ruse's admission that there are borderline cases with "no clean demarcation between science and non-science" hardly bolsters his case, in spite of the fact he says this admission is no argument against MN.[1] It is difficult to see how this admission does not hurt his case. If there are borderline cases, then there is simply no clear demarcation. Also, Ruse's response to Plantinga that his initial comments that science, by definition, is methodologically natural, were not an attempt at an analytical definition but lexical definition is simply odd. Does Ruse really believe that the general meaning or dictionary definition of "science" is what is going to settle this dispute?

A bigger problem for Ruse is the fact that it is not unusual for scientific studies to assess something other than what is considered strictly natural. For example, in 2006 a study was done on the effects of intercessory prayer for cardiac bypass patients.[2] While the study was unable to show that third-party prayer had any demonstrable effects on the patients, there is no doubt this was a "scientific" study even though it dealt with the supernatural. The fact is, scientific studies are regularly done which could possibly yield results that at least point toward the existence of the supernatural or paranormal. According to Ruse and others, however, studies of this sort should be considered non-science or

---

*Grunbaum* (ed. Robert Sonné Cohen and Larry Laudan; Boston: Reidel, 1983), 111-12. Emphasis Laudan's.

[1] Ruse, "Methodological Naturalism Under Attack," 49.

[2] Herbert Benson *et al.*, "Study of the Therapeutic Effects of Intercessory Prayer (STEP) in Cardiac Bypass Patients: A Multicenter Randomized Trial of Uncertainty and Certainty of Receiving Intercessory Prayer," *American Heart Journal* 151.4 (2006): 934-42.

pseudo-science for they attempt to say something about that which is not purely natural. If Ruse is right and that, by definition, science can only deal with natural matters, then the study of intercessory prayer was nonsensical from scientific standards. Or worse, the study was actually a bit of circular reasoning, for the only possible explanation of intercessory prayer would have to be, *a priori*, a natural explanation.

If arguments that science must be, by definition, methodologically natural fail, what are we to make of Plantinga's call for an Augustinian science? Can we have two kinds of science, one that is natural for the sake of neutrality and one that takes into account claims of the supernatural? While this does seem problematic on a practical level there does not seem to be any inherent reason why science cannot engage the supernatural. Ruse argues that there is no way to determine if Augustinian science is good science, but what would be his criteria? Refutations of supernatural science seem to fall back on versions of the argument that science is natural by definition. Practical concerns aside (such as the allocation of grant monies), difficulties in determining whether Augustinian science is as good as mainstream science (or as Ruse calls it, "science without qualification") hardly seem like a reason to deny it the status of science altogether. There seem to be disputes regularly over what kind of scientific explanation best suits the phenomenon. For example, scientists disagree whether homosexuality can best be explained by chemical, biological, or social factors. While there is no agreement as to which of these is clearly the better explanation, the response that only one can be good science seems problematic at best.

Pennock is most likely correct that theological disputes would keep Augustinian scientists from being unified. But again, is this a good reason for science to be inherently natural? As stated previously, this is a matter of practical issues and not whether science is inherently natural. Pennock's argument that to "apply natural knowledge to understand supernatural powers, then, by definition, they would not be supernatural" also has problems. Like Ruse's arguments, this seems to be a bit circular. For example, according to Maarten Boundry, Stefaan Blancke, and Johan Braeckman, we could imagine a situation in which Intelligent Design (ID) theorists provide firm and unambiguous evidence for ID behind biological complexity, such that even our metaphysics were challenged. If this were so, then, by Pennock's argument, he would have to reply by saying something like, "You see, now we have *scientific* proof for Intelligent Design. By definition, that means that we have to do with a *natural* phenomenon. Thus, I was right after all, supernatural causes and forces have no place in science."[1] But how could Pennock ever be refuted if each (supposed) scientific explanation of the supernatural meant that the supernatural was actually natural? Also, Pennock's argument here seems more theological than philosophical. Pennock is essentially claiming that if there is a supernatural being responsible for the universe, this creator would be unable to interact with the universe in any kind of supernatural manner. Any sort of supernatural intervention would have to be, by

---

[1] Maarten Boundry, Stefaan Blancke, and Johan Braeckman, "How Not to Attack Intelligent Design Creationism: Philosophical Misconceptions About Methodological Naturalism," *Foundations of Science* 15.3 (2010): 232.

definition, natural intervention. But it is unclear how Pennock's theological claims and *a priori* arguments dismiss Plantinga's call for an Augustinian science. There are certainly practical issues but these alone do not refute Plantinga.[1]

What about Pope's claim that Plantinga's case for Augustinian science, based on the incompleteness of MN, fails to overturn MN as the basis of science? Pope is right that we cannot make a case for MN's demise simply because MN has not brought to a close every scientific mystery. But conversely, is Pope's argument—that because MN has been fruitful we have reason for the dismissal of Augustinian science—actually valid? This hardly seems reason enough. Many scientific theories have been fruitful and yet eventually replaced. Newtonian physics was highly fruitful in its explanatory power and yet it was eventually succeeded by Einsteinian physics. Just as Einsteinian physics has not totally replaced Newtonian, Plantinga does not seem to

---

[1] Another way to approach Plantinga's claim is to consider Elliott Sober's argument that science regularly deals with the supernatural in the form of numbers. Sober states that "natural" entities, events, and processes have spatio-temporal location, while supernatural do not. By this definition, numbers (not numerals) are supernatural. In this sense, scientists follow Platonism, for it is common to think of numbers as real. Furthermore, Sober argues that supernatural claims are testable. For example, it can rightly be claimed that there are an infinite amount of prime numbers and it could be claimed that the number of apples in the basket is prime. The infinite cannot be empirically investigated and yet the claim that the number of apples in the basket is prime is perfectly empirical. For more information see Elliott Sober, "Why Methodological Naturalism?" This is an unpublished paper available at http://philosophy.wisc.edu/sober/why%20methodological%20naturalism%20rome%20format%20june%202009.pdf (accessed 11/29/2010).

be calling for Augustinian science to bring about the end of Duhemian, but merely that Duhemian science (like Newtonian physics) may have significantly limited explanatory power in the future. It seems clear that Ruse, Pennock, and Pope have not shown that there is an inherent reason to think natural explanations must be the only legitimate form of scientific explanation.

What about the claim that the end of MN would amount to science stopping? Most responses to Plantinga seem to ignore his main point and simply reinforce the claim that without MN science would stop. But this is not really a response to Plantinga, who simply argues this is not a problem. First, Plantinga does not come out and say, "science must end at some point," even if that is what he might actually think. What he says is *if* it is does stop, so be it. His point is that there appears to be no reason to assume *a priori* that science, as methodologically natural, must find a natural explanation for everything. Given Plantinga's wider argument, his point is that only Duhemian science, not Augustinian, would stop. This argument simply reinforces his point that there is room for more than MN in scientific exploration.

Let's take another route with this. Arguments, which suggest that without MN science would stop, are basically arguing that all "good" or "successful" science would end. But is this what history shows? It was not unusual for scientists like Galileo, Brahe, and Newton to include God in their scientific explanations. Newton, for example, believed that our solar system's stability was due to God's intervention. Newton argued, "God intervened to smooth down perturbations resulting from gravitational interactions among planets."[1] Certainly evoking the supernatural can be a science stopper but it need not always be a science stopper. Is it any more of a problem, theoretically, to say, "God did it" than to say, "evolution did it"? We would only have a science stopper if the scientist left it at that, if the final explanation was "God" or even "evolution." The task is to explain how God, evolution, or whatever, is responsible for the phenomenon. The issue of violating MN seems to be more of a statement about people than a necessary principle concerning MN. The argument seems to say that people would stop doing science, but there is no principle that says we would have to stop doing science.[2] While it could be suggested that if those who do stop, would simply work a bit further, they would most likely find a natural explanation for the phenomenon, this does not equate to all science necessarily stopping without the employment of MN. Neither does this explanation of why science must, in principle, never stop.

**Conclusion**

In this chapter I have summarized Plantinga's three philosophical arguments for why MN is not an inherent principle in the meaning of science or its theories and explanations. I have also briefly chronicled several responses to Plantinga's arguments by various philosophers of science. It is my opinion that each of these responses to Plantinga fails to demonstrate why his

---

[1] David B. Wilson, *Seeking Nature's Logic: Natural Philosophy in the Scottish Enlightenment* (University Park: Pennsylvania State University Press, 2009), 26.

[2] Sober, "Why Methodological Naturalism?" 14.

arguments are invalid and that MN is a necessary aspect of science.

It seems that the key issue in Plantinga's argument is that science need not be inherently—by definition—methodologically natural. If this is the case, then there is no inherent reason to not allow for Augustinian science. That is, there is no natural reason why science must always be methodologically natural. While there seem to be more practical reasons for supporting MN, advocates of MN and critics of Plantinga appear to be overly focused on *a priori* kinds of arguments. In other words, most of the arguments found in response to Plantinga suggest that Augustinian science is problematic or that empirical investigation must limit itself to the natural, simply because science is inherently naturalistic. But these are only more sophisticated versions of "science, by definition, is methodologically natural." I believe Plantinga has shown why arguments for MN, as an inherent scientific principle, ultimately fail.

In the end I am left to wonder whether the question of demarcation will ever be resolved. As shown in this chapter, border disputes can be rather heated at times. The demarcation of *science*, however, remains at once vastly important and yet, quite nearly impossible.

# *God, Miracles, Creation, Evil, and Statistical Natural Laws*

*Rem B. Edwards*

## Statistical Natural Laws

Consider the following metaphysical implications of the now generally accepted concept of natural laws as statistical.

1. *Before or without the existence of any actual entities, there are no actual laws of nature, only abstract possible laws for possible worlds or "cosmic epochs," as Whitehead would say.*

Alfred North Whitehead located all *possible* universals, qualities, relations, laws, and concrete individuals within the "primordial nature of God." He insisted that there are no novel eternal objects, thus no truly novel possible individuals, patterns, qualities, relations, or combinations of such. I disagree with Whitehead about this—as does twentieth century American philosopher Charles Hartshorne. There would be no difference in the nature or value of actual and possible worlds if this were so. Hartshorne thought, and I agree, that there are no complex eternal objects for concrete individuals like you and me; only the most abstract possibilities or repeatable "objects" are "eternal." God creates more definite possibilities as needed for their actualization. Calling *created* possibilities "*eternal*" would be self-contradictory, so I will refer to "possibilities," not to "eternal objects."[1] Possible laws of nature are the general formal patterns of relations between actual entities in one or more possible worlds. As Whitehead defined them for our world:

> The laws of nature are forms of activity which happen to prevail within the vast epoch of activity which we dimly discern.... There is no necessity in their nature.[2]

For this paper, I will also use "actualities" instead of Whitehead's "actual occasions."[3] By "actualities," I include existing events plus more enduring entities or structures like Whitehead's "regular trains of waves, individual electrons, protons, individual molecules, societies of molecules such as inorganic bodies, living cells, and societies of cells such as vegetable and animal bodies."[4] I will also include actualized repeatable properties (qualities and relations)—usually called "universals." I will not follow Whitehead in calling them "eternal objects" but will instead refer to "universals" or "properties."

2. *The actual laws of nature are created by the properties, dispositions, and habits of actual entities.*

---

[1] For a more complete critical discussion of Whitehead and Hartshorne on "eternal objects," see: Rem B. Edwards, "Whitehead's Theistic Metaphysics and Axiology," *Process Studies* 45.1 (2016): 8-10.

[2] Alfred North Whitehead, *Modes of Thought* (New York: Free Press, 1968), 87.

[3] I explain my doubts about "actual occasions" in Rem B. Edwards, *What Caused the Big Bang?* (Amsterdam: Rodopi, 2001), 244-53; and Rem B. Edwards, "The Human Self: An Actual Entity or a Society?" *Process Studies* 5 (1975): 195-203.

[4] Alfred North Whitehead, *Process and Reality* (Corrected ed.; New York: Free Press, 1978), 98.

The statistical laws prevailing in our universe are not directly "imposed" from above, outside, or before our universe. Instead, they are created by actualities within our universe.

Statistics do not apply to individuals, only to large sets, collections, or classes of similar actualities sharing similar properties (qualities and relations, habits, or dispositions). "All electrons are very similar to each other,"[1] said Whitehead. So are all things belonging to our conceptual classifications. Whitehead repeatedly affirmed statistical natural laws—as derived from the habits, actions, properties, or "characters" of similar actualities. Consider these examples:

> This doctrine, that order is a social product, appears in modern science as the statistical theory of the laws of nature....[2]
>
> Statistics tell you nothing about the future unless you make the assumption of the permanence of statistical form.... There is no valid inference from mere possibility to matter of fact.[3]
>
> The laws of nature are the outcome of the characters of the entities which we find in nature. The entities being what they are, the laws must be what they are; and conversely the entities follow from the laws.[4]
>
> Thus in a society, the members can only exist by reason of the laws which dominate the society,

---

[1] Alfred North Whitehead, *Science and the Modern World* (New York: Free Press, 1967), 109.

[2] Whitehead, *Process and Reality*, 92. See also 207.

[3] Alfred North Whitehead, *Adventures of Ideas* (New York: Free Press, 1933), 126.

[4] Alfred North Whitehead, *The Concept of Nature* (Cambridge: The University Press, 1971), 142.

> and the laws only come into being by reason of the analogous characters of the members of the society.[1]
>
> But the laws of nature are derived from the characters of the societies dominating the environment.[2]

Whitehead, and theists influenced by him, distinguish between natural laws as "imposed" from above, versus "immanently derived" laws, as required by the statistical theory. We still want to say something closely resembling "imposed," as in Whitehead's own words, "conversely entities follow from the laws," and "the members can only exist by reason of the laws." Here we must be very careful with our language, perhaps even more careful than Whitehead. We theists are accustomed to thinking that God gives the universe its laws, so we may easily lapse into language incompatible with statistical laws as "the outcome of the characters of the entities which we find in nature."[3] So what exactly do we want to say, and to avoid saying? "Imposed" natural laws are generally associated with predestination and rigid determinism, with "Law imposed by the will of inflexible Allah."[4] Clearly, we process theists do not want to say that. Statistical laws are not rigid or imposed, and we strongly prefer "persuasion" to "force."

Still, we want to affirm that somehow God "gives" the universe its basic order, its formal statistical patterns or laws. Whitehead said, "The quantum is that standpoint in the extensive continuum which is

---

1 Whitehead, *Process and Reality*, 91.

2 Whitehead, *Process and Reality*, 205.

3 Whitehead, *Concept of Nature*, 142.

4 Whitehead, *Adventures of Ideas*, 135.

consonant with the subjective aim in its original derivation from God. Here 'God' is that actuality in the world, in virtue of which there is physical 'law.'"[1] The entities to which God gives subjective aims already exist, are already actual and habituated, already have their own subjective forms, and must choose whether or not to accept novel aims. Most but not all process theologians reject creation *ex nihilo* and affirm the creation of our universe and its ingredient actualities out of the ashes or chaos of antecedent oscillating universes—going back to infinity.[2] This tension leaves process theists with the problem of what God does for the world, if anything at all, since the "stuff" of all universes co-exists with God everlastingly. Yes, God "persuades" by providing novel possibilities or subjective aims, and God's memory "saves" the actualities of the world as they perish in time, but what else? One answer is that God gives the world its natural laws. Hartshorne wrote of "laws of nature (which, some of us believe, are divinely decided and sustained)" and that "it is no small thing to give the world sufficient orderliness to make it possible for free creatures... to adapt to one another essentially harmoniously."[3] We clearly reject rigidly deterministic laws, but in what sense can God "decide" or "give" the world its lawful order without "imposition"? What exactly does God do?

---

[1] Whitehead, *Process and Reality*, 283.

[2] David Griffin, *Panentheism and Scientific Naturalism: Rethinking Evil, Morality, Religious Experience, Religious Pluralism, and the Academic Study of Religion* (Claremont: Process Century Press, 2014), 24-26, 100-09.

[3] Charles Hartshorne, *Omnipotence and other Theological Mistakes* (Albany: State University of New York Press, 1984), 18, 118.

The most carefully worded process answer is something like this. God decides *what kind of actualities* will exist within any given cosmic epoch or universe. That is, God imposes their *characters,* their most elemental properties. According to Lucien Price, Whitehead once said, "Why talk about the 'laws of nature' when what we mean is the characteristic behavior of phenomena within certain limits at a given stage of development in a given epoch—so far as these can be ascertained?"[1] Thus, in selecting *the relatively enduring properties* of the most fundamental *kinds of entities* in the universe, God indirectly imposed their basic *habitual activities.* The habits of enduring objects like atoms, molecules, cells, etc., issue partly from their given characters or properties, partly from their God-given ongoing novel aims, and partly from their own choices that can cumulatively create new habits. *Natural laws* as we know them are then *"derived from"* the characters, choices, activities, and habits of large quantities of similar actualities. They are the *formal numerable patterns of* such habits in mass. Thus, in choosing the elemental characters or properties of the basic actualities of the universe, God indirectly, not directly, "gives" the worlds its laws. A pure imposition theory says that God installed natural laws as such directly, quite apart from a universe's actualities. A statistically immanent theory says this happened only indirectly. Theistically, their "character" is God-given, but their actions jointly create their own formal statistical patterns or laws. We must avoid theological language that insinuates direct and separate instillation of rigid

[1] Lucien Price, *Dialogues of Alfred North Whitehead* (New York: Mentor Books, 1956), 278.

laws of nature by God or by anything else. By "laws of nature," we really mean, "patterns of activity derived from masses of similar actualities."

*3. The laws of nature are not efficient causes of anything; they are only formal causes. They do not forcefully make things happen or prevent anything from happening.*

Formal causes are universals. They are the patterns of the processes and interactions of very large numbers of concrete actualities and their inherent repeatable qualities and relations. They exist only within, between, and because of those actualities. They are derived or abstracted from those actualities. They do not actively cause those actualities. Actualities actively cause them. Formal causes have no force, energy, or power of their own; only efficient causes do. Because "efficacy" is so closely associated with "cause," it can be misleading to call them "causes" at all. So why do so?

Formalities of every description exist only within and because of actualities. Laws and other repeatable properties are effects, not efficient causes (hereafter "e-causes"). Some atheistic "scientific" cosmologists today contend that the Big Bang that originated our universe was originally caused, not by God, but by quantum fluctuations in absolute nothingness, as allowed or required by pre-existing laws of quantum physics, also located in absolute nothingness. I call this "Big Accident Cosmology" and devote a critical chapter to it in my *What Caused the Big Bang?*[1] There are many difficulties with this theory. Three obvious ones are these. First, these "scientists" are no longer doing empirical science. They are doing non-empirical, *a priori*, blind-faith,

---

[1] Edwards, *What Caused the Big Bang?* 163-78.

current-fad, "atheistic theology," to coin an accurate phrase that they would find obnoxious. Second, they are talking nonsense, for there is always "something" like pre-existing quantum laws in their "nothingness." Third, they are disregarding an empirical truth confirmed 100 percent of the time by experience: universals always exist within concrete individuals or actualities, never in complete isolation from them. Aristotle was right about that, as Whitehead agreed.[1]

This third point provides an empirical basis for Whitehead's "ontological principle" which affirms:

> [T]he notion of "power" I transformed into the principle that the reasons for things are always to be found in the composite nature of definite actual entities—in the nature of God for reasons of the highest absoluteness, and in the nature of definite temporal actualities for reasons which refer to a particular environment. The ontological principle can be summarized as: no actual entity, then no reason.[2]

Aristotle's final (teleological), efficient, and formal causes are all embraced by the ontological principle, especially the first two:

> It could be termed the 'principle of efficient and final causation.' This ontological principle means that actual entities are the only *reasons*; so that to search for a *reason* is to search for one or more actual entities.[3]

No quantum laws can exist and fluctuate in a nothingness in which there are no actual entities whatsoever. Experience, not just Whitehead's authority,

---

[1] Whitehead, *Process and Reality*, 40.
[2] Whitehead, *Process and Reality*, 19.
[3] Whitehead, *Process and Reality*, 24.

confirms this 100 percent of the time. "Everything must be somewhere; and... 'somewhere' means 'some actual entity.'"[1] If something transcendent explains our universe, a transcendent God is a much better answer than "nothingness" or "laws within nothingness," which are something after all.

Whitehead agreed with Plato that "being" (or "actuality") is power.[2] Abstract formalities like laws, universals, and possibilities do not have or exert any power, force, or energy of their own. They are not and cannot be the efficient causes (e-causes) of anything. They do not explain actualities; actualities explain them. Only actualities manifest agency and e-causation. Possible but not yet actual laws of nature do not predict, create, or produce, any actual entities, events, or universes. They cause nothing, explain nothing, predict nothing, resist nothing, and prohibit nothing. They place no limitations on anything or anyone, including God.

*4. The laws of nature evolve and change as the choices and habits of actual entities within the world evolve and change.*

Statistical natural laws are not all "given" or "fixed" at or before "the beginning" (at the Big Bang, we might say). They evolve because actualities evolve. Big Bang cosmologists trace the origin and evolution of physical entities like subatomic particles, photons, electrons, waves, fields, photons, atoms, molecules, aggregates, living things, etc., from the primordial soup of pure energy of the Big Bang itself. No pre-existent *laws* as such determined the nature and habits of anything; rather, such things evolved their own laws. Natural laws

---

[1] Whitehead, *Process and Reality*, 46.

[2] Whitehead, *Adventures of Ideas*, 119-20, 129.

merely sum up the general patterns of their habitual behaviors and interactions in mass. Before electrons, atoms, molecules, and living things emerged, there were no laws of physics, chemistry, biology, etc. Neither they nor their laws were instantaneously present in the Big Bang. Their laws emerged only as these actualities emerged. Divine "persuasion" influenced and encouraged their emergence. The view that atemporal pre-existent laws caused their emergence has the cart before the horse.

Because their formal laws emerged from them, not vice versa, individual natures, essences, rules, or relatively enduring properties are never totally fixed by their efficient or formal causal antecedents. Degrees of self-creation and natural law creation are very real. Process metaphysics make this very clear.

*5. Statistical laws of nature do not tell us what any particular actual entity is doing or must do. They tell us only what large masses of similar actual entities are doing, have done, or will do, on average.*

The laws of nature known to us describe or express only statistical averages. They do not dictate or describe what happens to each or any individual covered by such laws. They do not prescribe, determine, coerce, or predict what any given individual must or will actually do. This is conspicuously true in some cases. Statistical laws pertaining to radioactive elements like radon and uranium recognize that large quantities of their molecules will spontaneously discharge a predictable percentage of their electrons within a given period of time. However, exactly which particular molecules will do this is completely undetermined and undeterminable. Heisenberg showed that the positions and velocities of subatomic particles are not simultaneously determined or determinable. In evolutionary biology, no laws predict or explain the detailed novelties of "punctuated equilibrium." At the human level, insurance companies can predict quite accurately how many people will get sick or die in a given period of time, but they cannot identify any specific individuals who will do so. Statistical actuarial laws say only that on average, a relatively definite (but otherwise unidentifiable) number will die.

All statistical natural laws posit only high probabilities, not absolute certainties. They leave room for a few of the instances they cover to be very atypical without affecting, changing, or violating the statistical averages. Some atypical and unpredictable instances might be miracles, *if* there are any. When averaged in with the vast quantities of events covered by statistical "laws of nature," an occasional old-fashioned divine miracle or "anomaly" would not make any noticeable difference at all in those statistics, so in what sense could such miracles be said to "violate" them?

**God and Miracles**

Does God work miracles, and if so, why do so many evils exist in the world? Such problems are immensely complicated. What do we mean by "God," "miracles," and "evils?" Perhaps nature's laws as statistical can shed some interesting and meaningful light on such things.

Traditionally, "miracles" meant that God's transcendent e-causation occasionally violates, suspends, or overrides the absolute laws of nature, but since these laws are only statistical, not absolute, all relevant issues must carefully reconsidered.

In traditional debates on "miracles" as violations of natural laws," both sides presupposed that natural laws are e-causes that force some things to happen and prevent others from happening. However, we now know that statistical natural laws are the effects, not the causes, of any actual happenings. Actual happenings create statistical natural laws, not vice versa. The traditional understanding of miracles as "violations" of natural laws meant that God sporadically causes odd and unexpected things to happen by exercising transcendent e-causation that overcomes the e-causation of the inviolable laws of nature. "Scientific" or "naturalistic" minds refuse to accept this. However, statistical natural laws also possess, exert, and resist no force. They are purely formal, not efficient causes. Only actual forces can "violate" (overpower) or be "violated by" by other actual forces. Atheists say that since God does not exist, God has and uses no power. Yet, statistical natural laws have no power either. This calls for further consideration.

"Miracle" has several meanings. In addition to violations of natural laws, miracles may also be things that happen which are surprising, unexpected, and unpredictable. In this sense, miracles that break no laws happen every day. Even at a deep physical level, many odd and unpredictable things happen in our universe. If and when enough odd but similar things happen, they become commonplace; their collective habits become the laws of nature. When electrons and photons *first* emerged from the original primordial soup of pure energy, there were no natural statistical laws for them. No natural laws existed to determine them in advance or to predict their properties or propensities. The whole course of physical, chemical, and biological emergence illustrates the emergence of new laws of nature. So does the whole course of biological evolution, evolutionary psychology, and human sociology and history, for which there are hardly any natural laws at all, if any. One does not have to be a process thinker to hold such views.

When we wonder if God works miracles, do we want to know if God somehow causes surprising and unpredictable events to happen, or do we want to know something more? Miracles as astonishing and unpredictable events regularly happen under God's lure or inspiration, so process theists believe. Providing actual entities with novel "actual aims" involves more than God's final/formal causation. Acting directly on actual entities to give them novel actual aims also involves a degree of efficient causation. Whitehead acknowledged this in his own technical terminology when discussing "hybrid physical feelings," understood as forms and aims derived from some active "physical" source such as God. He proclaimed, "All conceptual feelings are derived from physical feelings," even those that temporal actual entities derive from God.[1] So, even God's "persuasion" involves transcendent efficient as well as final and formal causation. If God actively gives actualities novel aims every moment, very small e-causation miracles occur constantly. God's usual "miracles" are almost imperceptible and not very surprising. Theists are most interested in miracles that are perceptibly large, noticeable, and quite astonishing. Small or large, God injects transcendent efficient causal energy and novel possibilities into the immanent world without "violating" (overpowering) the statistical laws of nature—which have no energy or e-causality of their own. If God uses e-power to initiate unexpected events

[1] See Whitehead, *Process and Reality*, 246-47.

or minor "fluctuations" within events, the use of that power would never "violate" the *powers* of the laws of nature. There are none.

Saying that God gave our universe its natural laws is only a roundabout way of saying that God gave certain relatively enduring properties, powers, qualities, and relations to the universe's enduring actualities. To affect nature's laws indirectly, God must act directly and creatively on substantial numbers of its actualities, not on its laws. God does this in process theism by inspiring or infusing actualities with novel possibilities or aims for further self-development.

### God, Power, and Evil

The really troublesome concern about God's e-causation power is ethical or axiological, not metaphysical. If God has the e-power to do surprising and unpredictable things without "violating" statistical natural laws, why doesn't God surprise us much more often by preventing both humanly initiated moral evils *and* naturally occurring evils like diseases, deformities, injuries, catastrophic epidemics, tornados, hurricanes, earthquakes, and tsunamis? Perhaps God does not prevent *moral* evils, the terrible things we freely decide to do to one another, because doing so would negate our freedom, creativity, and self-control. But *natural* evils are another story.

We can identify at least four process "solutions" to the problem of "theodicy," the attempt to reconcile God's goodness and power with the presence of evils in the world. Clearly, morally good human parents would *actually do something*, where possible, to prevent their children from being harmed by both moral and natural causal agents, but God clearly does not do so. Why not?

Is God less good than loving human parents? How can God be morally good and not prevent all such harms? At least four process theodicies try to explain how God can be supremely good, yet fail to prevent evils or harms.

1. God simply *lacks transcendent e-causal power* to prevent evils, work traditional miracles, and create universes out of nothing. Also, some additional necessary and independent metaphysical principle like *creativity* prevents God from doing so.

2. God *has the transcendent e-power* to prevent evils, work miracles, and create universes out of nothing, but God usually limits this power *voluntarily* so some if not all creatures can be free, creative, and morally virtuous and responsible. However, God occasionally works e-causation miracles and did create our universe out of nothing.

3. God does not prevent evils because God's *necessary* (not voluntary) love and moral goodness does not allow God to override the freedom and creativity of creatures. God also *lacks the power* to do so, that is, to work traditional efficient-causation miracles (hereafter "e-miracles"), and to create universes out of nothing. God nevertheless works miracles-by-persuasion (hereafter "p-miracles").

4. God *has the power* to prevent evils, to work e-miracles, and to create universes out of nothing, but God does not work e-miracles because doing so would be immoral, that is, *unjust* and unloving. God does work nontraditional p-miracles, however, and God used transcendent e-causation to create our universe out of nothing.

The numbers below correspond to the four positions outlined in the preceding paragraphs.

1. David Ray Griffin offers and vigorously defends the first process theodicy. His view, expressed in many publications, is that God is not the e-cause of anything. God is only a final and formal cause. God lacks the power to work e-miracles to prevent harms. God also lacks the power to create universes out of nothing, for if God had it, God would also have and inevitably would use this power to work e-miracles, which does not happen. God affects the world only by persuasion, never by working e-miracles that interfere with or overpower nature's laws. Griffin completely renounces God's e-power (traditional omnipotence) in order to save God's goodness.[1]

In addition, Griffin maintains, an independent necessary metaphysical principle, *creativity*, absolutely prevents God from acting with direct causal efficacy to prevent evils. Necessarily, all actualities are partly self-creative, and God lacks the power to interfere with individual or cumulative creature-creativity in order to avert harmful consequences.[2]

2. Process theists come in many varieties. In addition to mainstream process theologians like David Griffin and John B. Cobb, Jr., many others, who prefer to call themselves Relational or Open theologians, are also process or temporalistic theists who attribute change, process, or temporality to God. They generally affirm that God has transcendent e-causal power to create our universe out of nothing, and to work occasional e-causation miracles that "violate" the laws of nature. God

---

[1] Griffin, *Panentheism and Scientific Naturalism*, 117-27.

[2] Griffin, *Panentheism and Scientific Naturalism*, 90-91, 118-22, 124, and especially 255-56. For my critique of this, see Edwards, "Whitehead's Theistic Metaphysics and Axiology," 21-23.

did in fact create the universe *ex nihilo*, and God does work such miracles occasionally, they think. However, God usually does not work miracles to prevent evils because God *voluntarily* limits God's own power so creatures can be free, creative, and morally responsible.[1]

3. Thomas Jay Oord is the best representative of the third and newest process theodicy, though I find some ambiguity in his thinking about God's e-causality. His most recent book, *The Uncontrolling Love of God*, is his most thorough and convincing presentation of the view that God *necessarily* refrains from e-causation miracles, not just *voluntarily*. Oord regards himself as an Open and Relational Theist, but he disagrees with those who hold that God refrains *voluntarily* from preventing evils so that creatures can be free, creative, and morally virtuous and responsible. Instead, God *necessarily* refrains because of God's necessary, involuntary, uncontrolling love and moral goodness. Refraining, or not, from e-miracles is never an optional matter of voluntary choice for a loving God. God "necessarily provides freedom, agency, self-organization and regularity to creation" and cannot withdraw or override it.[2] This is because of God's primordial, necessary, and involuntary moral attributes of love and goodness,[3] not because creativity as an

---

[1] Excellent critical discussions of these "voluntary self-limitation" theists can be found in many of Tom Oord's books. See Thomas J. Oord, *Defining Love* (Grand Rapids: Brazos Press, 2010), 147-212; *The Nature of Love: A Theology* (St. Louis: Chalice Press, 2010), 85-157; *The Uncontrolling Love of God: An Open and Relational Account of Providence* (Downers Grove: InterVarsity Press, 2015), 107-49.

[2] Oord, *The Uncontrolling Love of God*, 169.

[3] Oord, *The Uncontrolling Love of God*, 170-71.

independent metaphysical principle, or because anything outside of God, "imposes limitations."[1]

I find some ambiguities in Oord's thinking about whether or not God has or lacks the power to work e-causation miracles and create *ex nihilo*. On the one hand, he says, "But my theory is essentially neutral on the issue of *creatio ex nihilo*,"[2] that God as a spirit "exerts efficient causation,"[3] and that God has the power to give power to others.[4] On the other hand, he clearly affirms that "*creatio ex nihilo* should be abandoned,"[5] partly because it is an extra-biblical doctrine, added later, partly because it ranks God's power over his love, etc. He also seems to accept Griffin's often repeated argument that if God had the efficient causal power to create *ex nihilo*, God would also have the power to work law-violating efficient causation miracles, and God would inevitably do so,[6] which God does not do, so God must lack e-causal power. However, Oord believes, God definitely does work large-scale surprising-to-us miracles by persuasion—like those in the New Testament—by coaxing the relevant atoms, molecules, cells, organs, organisms, and physical processes to do very surprising things coordinately.[7] Oord may hold that God's e-causation is never sufficiently strong to override creaturely freedom and self-determination, but is this a metaphysical or a moral necessity for God, or somehow both?

---

1 Oord, *Nature of Love*, 125.

2 Oord, *Uncontrolling Love of God*, 149, n. 64.

3 Oord, *Defining Love*, 194, n. 54.

4 Oord, *Defining Love*, 210-11.

5 Oord, *Defining Love*, 207, n. 79.

6 Oord, *Defining Love*, 143, 106-07.

7 Oord, *Uncontrolling Love of God*, Ch. 8.

4. The fourth view is that God has sufficient e-causal power to prevent evils, but God does not do so from love, justice, and moral goodness—with special emphasis on *justice or fairness*. This is my own view. It is closer to Oord's than to any of the others. It might be identical with Oord's if certain ambiguities in his position were removed, and if he were more open to the possibility of creation *ex nihilo*. I disagree with Griffin because, like most Open and Relational theists, I think that God does indeed have sufficient e-causal power to create universes *ex nihilo*. I also think that creativity is entirely at God's disposal and is not an independent metaphysical principle that imposes limitations on God. I side with Oord against the "voluntary self-limitation" position, and hold with him that not preventing harms is *necessitated* by God's uncontrolling love, justice, and moral goodness.

However, my view emphasizes *justice* and other positive moral goods in addition to human freedom, creativity, self-determination, and responsibility. More than these are at stake in matters of theodicy, and Oord would agree in his own way.[1] My best account of what I believe to be a workable process theodicy is in my book, *What Caused the Big Bang?*[2] There I explain that there is no single "silver bullet" that reconciles God's power with God's goodness. Only the cumulative force of a number of considerations, outlined next, will work. Each is explained in some detail in my book.

1. The *free-will defense*, according to which God in his goodness, not from metaphysical necessity, gives creativity, self-determination, and degrees of "free will"

---

[1] Oord, *Uncontrolling Love of God*, 169, n. 41.

[2] Edwards, *What Caused the Big Bang?* 295-310.

to all creatures throughout all creation, not just to human beings. Moral evil results from our abuse of this gift, and much natural evil results unintended from collective creative decisions made down through the depths of nature.

2. The *soul-making defense*, heavily emphasized by John Hick, according to which it would be impossible for human beings to have, develop, or exercise many, if any, moral and spiritual virtues in a universe lacking all dangers and evils.

3. The great *utility of law and order*, which enables us to partly predict and control much of the future, including the desirable and undesirables consequences of our own actions.

4. The *inevitable conflict of good with good* in any rich and complicated universe, a theme heavily emphasized by Charles Hartshorne.

5. The great *consolation* derived from knowing that God suffers with us in our suffering.

6. Possible *compensation* in "life after death."

7. To these, I will shortly add a seventh: *God's justice or fairness to all*. I missed this earlier in my book, but now regard it as an essential addition to the cumulative effective of a number of elements in a workable process theodicy.

More values and disvalues must be considered when doing theodicy than whether or not a good and powerful God could or would prevent harms through e-causation. My approach considers both preventing harms and sustaining many positive values. God's love, providence, and moral goodness could be expressed in all of the above ways, (necessarily so—given Oord's recent influence upon my thinking). No one of these alone is sufficient for a plausible process theodicy, but collectively they seem to me to be.

Thomas Jay Oord holds that the only way God *will* influence or change the world is through persuasion, that is, through a combination of final and formal causation. He seems to differ from Griffin in thinking that this is the only way God *can* do this. Does God actually have sufficient e-causal power to influence or change the world more directly, other than by persuasion alone? Most Relational and Open process theists would say, "Yes," and so do I, but with the qualification that God *has* the power to work e-miracles, but *does not use* it for moral reasons, that is, because God is *fair or just*. Can this compromise position can be developed in a way that would make it plausible to all process theists? Please consider the following:

In my *What Caused the Big Bang?* I allowed at the time that God might occasionally work e-miracles that forcefully override natural laws.[1] I now see, however, that because natural laws neither force nor prevent anything, they cannot be forcefully "violated." God can still influence and change individual properties, qualities, and relations, but how? Part of the answer surely involves God's luring, inspiring, and exercising final/formal causation. In this way, massive microscopic creative changes that noticeably alter the macroscopic patterns of activity that we call the statistical laws of nature usually occur very, very slowly. In dramatic cases like the miracles of the Bible, however, Oord's rather swift miracles-as-persuasions (hereafter, "p-miracles") of massive numbers of relatively localized actualities might explain how God does things that greatly surprise us,

---

[1] Edwards, *What Caused the Big Bang?* 256-57.

especially natural scientists and naturalistic atheists. Even here, though, we can still wonder why God doesn't work more p-miracles (miracles-by-persuasion) to prevent terrible harms. Perhaps this is because most partly self-creative but already habituated actualities refuse to cooperate in making rapid cumulative physical changes, especially with respect to creating novel physical obstacles that would block harms.

I end up where Oord does, with God as necessarily uncontrolling and non-overriding for moral reasons, though perhaps I give slightly more emphasis to justice, which he would also include within love. I agree with Oord that God may work dramatic p-miracles (though rare, unlikely, and not an easy task). Unlike Oord, I want to say unambiguously that God *has* sufficient e-causation power to work miracles, but God does not *use* it for reasons of *fairness*.

My own prior publications in philosophical theology consistently allowed for the possibility of rare e-miracles. I never ruled them out *a priori*, as do many process thinkers like David Griffin. Occasional miracles to prevent harms or simply to announce God's presence or guidance, either by e-causation or by persuasion (final/formal causation), still present serious theodicy problems. If God *occasionally* works *either* e or p miracles to prevent pointless suffering, incapacitation, deformity, and premature death, why not *always*, as any loving parent would do? God is supposed to be much more loving, powerful, and knowledgeable than human parents.

Part of the answer is that a loving God would necessarily give freedom and not override it, as Oord insists, but there is more to it than that. In addition to *free will* and *soul making*, highly relevant also is the great *utility of law and order* that enables us to predict and control much of the future. To make a long story short, consider the immense instrumental goodness of having reliable "laws of nature," even if they are only formal statistical summaries of the habits of very large quantities of similar actualities like atoms, molecules, cells, and organisms.

Loving human parents are in no position to work e-miracles, that is, to make sudden and drastic localized changes that seem contrary to the general statistical laws of nature. But if God can work e-miracles, why does God not do so? Partly, this is because our dwelling in an orderly and predictable universe is itself a very great instrumental good for us and other living beings. The dependability of nature is worth the price of many of the evils that result from the general uniformities we abstract and conceptualize as natural statistical laws. Their advantages to us usually outweigh their disadvantages, but not always. Usually they work for us, but when they work against us, why doesn't a loving and e-powerful God *always* work e-miracles to "save us from all ills"? This is mainly because if we knew we could expect God to solve all our problems for us and save us miraculously every time we get into a jam, we would never develop into morally conscientious, virtuous, creative, and responsible persons.

In a universe of unfailing divine miracles, we would never develop *any* human virtues or seek *any* knowledge, all of which hinge on our own needs, curiosity, choices, efforts, actions, foresights, insights, growth, and maturity. Expecting miracles can be a way of avoiding personal responsibility. Soul-making reenters the picture unexpectedly at this point. Reliable laws of nature enable, promote, and demand it. Most nonhuman animals learn and generalize from experience, but even they would not learn, try, grow, and mature in their own more limited yet significant ways if God's e-miracles were universal. If e-miracles were universal, they would simply be the statistical laws of nature, and we would rely totally on them. But many valuable things would be absent from such a universe.

An *occasional* e- or p-miracle to prevent harms might not block all moral and spiritual growth and undermine all our efforts to take responsibility for our own lives and influences. The trouble is, a morally good and *just* God would *always* prevent *all* harms, as would morally good parents who have the knowledge and power to do so. Of course, such parents gradually relinquish control as their children mature. God's preventing *all* harms would completely overturn all the regularities of nature as we know them. Then we could and would not control the future course of our lives (as we now do within limits). This would undermine all our efforts and desires to control our own destinies and prevent harm to ourselves or to anyone else. Indeed, if we could always rely on God to solve all of problems for us, why bother with or care about anything? If God did everything for us, we would never do anything for ourselves. We would not be creative or moral beings at all. Can we even imagine living in such a universe?

But why insist on either the *universality* of e-miracles, or *none* at all? This is because a morally good God would be both loving *and just*—necessarily. Love inevitably includes justice or fairness, especially where more than one individual is involved. To take account of God's *justice*, (and to more definitively resolve problems of theodicy), I must now add the seventh highly significant consideration to the six explained in my Big

Bang book. I found this new-to-me argument in the third chapter of *The Predicament of Belief: Science, Philosophy, and Faith* by Philip Clayton and Steven Knapp.[1] To summarize briefly, if God were to intervene only selectively and occasionally to prevent harms to *a few* people (or animals) in danger, but *not to everyone* in danger, this would be incredibly *unjust or unfair*. Where only one out of a hundred people survive an airplane crash, some say that God deliberately and actively saved only the one, but, by implication, not everyone else. How horribly unfair this God would be to the ninety and nine others! Such an unjust God would be downright immoral, horribly so, definitely not that Supremely Good Reality than whom none better and more worshipful can be conceived. God's justice, as well as God's love, requires God either to work e-miracles to prevent all harms *always*, or *never*. We now understand how disastrous *always* would be. So, a supremely good, loving, and just God who *has* the power to work e-miracles would never *use* it, *because doing so selectively would be infinitely unfair or unjust to everyone not so favored*.

So where are we? The possibility of Oord's p-miracles, usually ineffective because the ability of concrete actualities to resist persuasion remains intact. But a necessarily moral, loving, and just God who deserves our supreme devotion would necessarily *have* but never *use* infinite e-power to work traditional miracles. Yet, God could still have and use such infinite power to create universes *ex nihilo*. Many Open and Relational process theists believe that without infinite e-

---

[1] Philip Clayton and Steven Knapp, *The Predicament of Belief: Science, Philosophy, and Faith* (Oxford: Oxford University Press, 2013), 44-68.

power, God would not be that Supremely Good Reality than whom none more admirable and worshipful can be conceived. I agree. God's power is not God's supreme perfection-making attribute, as it was in much of classical theology. Nevertheless, God's e-causal power has great perfection-making significance. Something highly desirable would be lacking in a Supercelestial Wimp who only nags but can never actually do anything. Infinite power not used for e-miracles would not be useless to God, who could and would use it to create universes to love. God's traditional omnipotence need not be rejected in order to explain why a morally good God would not "violate the laws of nature" to save us from all ills. God's justice definitively breaks the allegedly inevitable connection between e-miracles and e-creation presupposed by Griffin and others.

Even if unbiblical, there might still be good *philosophical* reasons for accepting creation *ex nihilo*, and for rejecting both atheistic explanations of the Big Bang and the alternative of an infinite number of antecedent, oscillating, God-influenced universes presupposed by process theists like Hartshorne, Griffin, Cobb, and Oord. I argue for this forcefully and at great length in my *What Caused the Big Bang?* There I also explain how and why process theists can and should affirm creation *ex nihilo*, and how God could be everlastingly creative of universes to love, yet still create our universe out of nothing. But all of that is beyond the scope of this paper.

Thus, I conclude (with justification, I think) that when God created our world out of nothing, there were no pre-existent individuals, independent metaphysical principles, or lawful regularities to "violate." There was only God in God's necessary and infinite goodness, love, justice, knowledge, wisdom, and power. "In the beginning," God had it all—and used it.

## *Science and the* Sensus Divinitatis

### *The Promise and Problem of the Natural Knowledge of God*

*Greg Cootsona*

*We believe in one God the Father Almighty, Maker of heaven and earth, of all things visible and invisible.*
The Nicene Creed

In the first article of the Nicene Creed, the Christian church confesses that God created the universe and thus humankind. God's creating this world implies that all human beings possess some natural knowledge and thus a yearning for God. Thus, the doctrine of creation situates the natural knowledge of God[1]—and more particularly, John Calvin's *sensus divinitatis*—in its appropriate context.

Before proceeding further, I need to emphasize the conviction that theology must *work* for the church. This ambiguous phrase is meant in a number of ways: First of all, theology must *work to make the church better.*

---

[1] The phrase "the natural knowledge of God" comes from chapter 2 of Wolfhart Pannenberg, "The Concept of God and the Question of Its Truth," *Systematic Theology* (trans. Geoffrey W. Bromiley; Grand Rapids: Eerdmans, 1991), 1:73ff. Pannenberg notes that older Protestant dogmatics differentiated the natural knowledge of God from revealed theology. For the purposes here, the term is more or less equivalent with *general revelation* (Pannenberg, *Systematic Theology* 1:73), as I will clarify below.

Naturally, the most significant representative for this position is Karl Barth. In his magisterial *Church Dogmatics* Barth explicitly inserted the German word *kirchliche* ("church" as an adjective) in the title to his dogmatics to demonstrate that theology "is a function of the Church."[1] Theology must *work* for the Church in that it makes the church a better place. It must also *work* in that theology should make sense of Christian life and practice. In this light, I will offer theological reflections that take in the insights of science, the necessities of the church, and the teachings of Scripture.

I cited Barth above with approval, but what he did not do sufficiently, and what I am seeking to do here, is to place theology in a "creative mutual interaction" with science, to use Robert J. Russell's phrase,[2] and set theological insights within the typology of a Lakatosian "research programme."[3] I am convinced that, in this interaction, theology must grasp, while at the same time not violate, the insights of science.[4] As John Polkinghorne rightly argues, a scientifically informed theology demonstrates that we are inherently motivated to believe what is truth and that our beliefs correspond

---

[1] Karl Barth, *Church Dogmatics* I/1: *The Doctrine of the Word of God* (trans. Geoffrey W. Bromiley; Edinburgh: T & T Clark, 1975), 3.

[2] Robert John Russell, *Time in Eternity: Pannenberg, Physics, and Eschatology in Creative Mutual Interaction* (Notre Dame: University of Notre Dame, 2012), 10-11; Nancey Murphy, *Theology for a Scientific Age* (Ithaca: Cornell University Press, 1993).

[3] In proposing a "research programme," I have in mind Imre Lakatos's theory of science. See especially, his chapter, "Falsification and the Methodology of Scientific Research Programmes," *Criticism and the Growth of Knowledge* (ed. Imre Lakatos and Alan Musgrave; Cambridge: Cambridge University, 1970), 91-106.

[4] See John Polkinghorne, *Science and Religion in Quest of Truth* (New Haven: Yale University, 2011).

to reality within the framework of critical realism. Simply stated, Polkinghorne argues that theology is "motivated belief."[1] To put it in more traditional language, the Book of Scripture and the Book of Nature have the same Author, and therefore do not contradict one another.

In this light, applying scientific insights to theology provides a test case for the natural knowledge of God, not least because some scientists argue that their disciplines can demonstrate God's existence or non-existence.[2] In addition, a natural knowledge of God might be demonstrated, or at least supported, by the insights of science. And finally, the scientific study of nature also flows from a commitment—whether explicit or not—that the world is rational and ordered, which historically has flowed from the confession that God created this world. In a lecture on the relationship between the Christian faith and modern science, the Nobel laureate Charles Towne summarized this connection: "For successful science of the type we know, we must have faith that the universe is governed by reliable laws and, further, that these laws can be discovered by human inquiry."[3]

I arrive then at my set of guiding questions: is there a natural knowledge of God? How does that relate to science? And what does this mean for the church?

---

[1] Polkinghorne, *Science and Religion*, 18. Cf., Justin Barrett, *Cognitive Science, Religion, and Theology: From Human Minds to Divine Minds* (Templeton Science and Religion Series; West Conshohocken: Templeton, 2011).

[2] Richard Dawkins infamous atheism comes immediately to mind as in *The God Delusion* (New York: Mariner, 2008).

[3] In Ted Peters, ed., *Science and Theology* (Boulder: Westview, 1999), 46.

My response will be in seven parts. First, I outline the natural knowledge of God and particularly John Calvin's *sensus divinitatis*. In the next two sections, I move to the natural knowledge of God in the Bible and subsequent theological tradition. In the fourth and fifth section, I move to science: first in the concept of beauty in both theology as well as scientific theory, and then to the specific contributions of the Cognitive Science of Religion (CSR) in understanding the natural knowledge of God. In the sixth section, I offer a theological critique of the *sensus divinitatis* and conclude, in my seventh section, with a Christological reconstruction of this concept.

**The Natural Knowledge of God and Calvin's *Sensus Divinitatis***

At times in my work as a pastor—and in response to this search for a reasonably intricate theology—I can hear someone reply: "Last weekend, I spent time in the mountains, gazing across a cool, still lake, listening to the wind through the trees. I was able to be silent. To be honest, I often find the Bible confusing, but in the quiet of nature, I directly encountered God. I learned more about God there than I ever do in a worship service. On Sunday mornings, I hear *about* God. There I actually touched my Creator."

In many ways, this natural knowledge of God is anti-ecclesial. It poses the question: Why do I need church when I have this direct experience through nature? And particularly, can I learn all I need to know about God through scientific analysis of the natural world? Indeed, why do I need a message from the pulpit when there are "sermons in stones" (to quote William Shakespeare)? From my pastoral experience—and, really, my experience generally—many people, religious or not, find an almost palpable presence of God in creation. Here, a few definitions will help. In theological language, we enter the realm of *general revelation*, where God speaks and acts in revelation—not *specially* through Scripture, but *generally* to all human beings though nature. In many ways, my reflections on the natural knowledge of God constitute a form of general revelation, which also implies God's benevolence toward all human beings, whether believer or not. As the Gospel of Matthew phrases it (Mt 5:45 NRSV): God "makes his sun rise on the evil and on the good, and sends rain on the righteous and unrighteous." Theologians contrast it—or complement it—with *special revelation*, God's particular acts and communication with the covenant people of Israel and the church.[1] The key point to emphasize is that in either general or special revelation, *God* is still the One revealing. God is the One who must speak in self-revelation. In this chapter, I will only briefly touch on a related area, *natural theology*—which takes the data of nature and seeks to build a theological system—and particularly what it means within the critical interaction of science and theology.

The natural knowledge of God is the intrinsic human capacity and openness for God. It is a "nonthematic knowledge of God" (to quote Wolfhart Pannenberg)[2]. It is necessary for a fuller, more robust

---

[1] Cf. Thomas Aquinas, *Summa Theologica* (Great Books; trans. Fathers of the English Dominican Province; Chicago: Encyclopedia Britannica, 1952), 2.2.2a.

[2] Pannenberg, *Systematic Theology*, vol. 1: 116. Cf. the language of Alvin Plantinga that this belief in God is "properly basic" and is therefore no more in need of epistemic verification than atheism. His emphasis falls on evidence or rationality for belief in

knowledge of God, but it is vague and therefore malleable and open to distortion. It derives from the nature of nature, as it were. As Alister McGrath phrases it: "there is an intrinsic capacity within the created order to disclose God."[1] This relationship arrives from bearing God's image, the *imago Dei* (which I will develop in the next section).[2] To use John Calvin's phrase, it is a *sensus divinitatis*, or "sense of the divine" (which I will also develop below). This *sensus divinitatis* provides a background for a more robust and articulated faith in God. It does not guarantee that we know God consciously.[3] It is certainly not a fully articulated form of Christian faith, but it is endemic to human life and therefore an important component toward building an ecclesial theology informed by science.

To be clear, natural knowledge of God poses a challenge, and I am responding to this challenge by formulating the proper, useful, and even necessary place

---

God, which is certainly important, but different from mine, which is whether God is known, in some way, to all human beings. See Alvin Plantinga and Nicholas Wolterstorff, eds., *Faith and Rationality: Reason and Belief in God* (Notre Dame: University of Notre Dame, 1983), 16-93. A more concise form of this argument can be found in *The Intellectuals Speak out about God* (ed. Roy Abraham Varghese; Chicago: Regnery Gateway, 1984), 185-201.

[1] Alister McGrath, *A Scientific Theology* (Grand Rapids: Eerdmans, 2001), 1:297.

[2] See also Alister McGrath, *Science and Religion: A New Introduction* (Oxford: Wiley-Blackwell, 2010), 184-89.

[3] Here I am rejecting the main components of "Reformed epistemology" presented by Alvin Plantinga (such as in *God and Other Minds* [Ithaca: Cornell University, 1990]), because, in my mind, it is minimally an inadequate elaboration on Calvin's ideas. For a summary of its problems, see Derek S. Jeffreys, "How Reformed Is Reformed Epistemology? Alvin Plantinga and Calvin's '*Sensus Divinitatis*,'" *Religious Studies* 33.4 (1997): 419-31.

for the awareness of God in nature and thus in ourselves (our reflection on nature and our understanding of our own desires), as well as what science has discovered about the natural world. Put with utmost economy of words: A natural awareness of divinity is necessary, but not sufficient, for our understanding of God. In this regard, I am steering a path alongside Calvin's *sensus divinitatis*, seeking to avoid Barth's abhorrence of "natural theology" and of Vatican I's rather overblown declaration that God can be known with certainty by the natural light of human reason from created things.[1]

**The Natural Knowledge of God in the Bible**

Genesis 1-3 set up four basic human relationships, which are carried throughout the biblical texts: with God, with other human beings, with ourselves (implied), and with the rest of creation (other animals, plants, and the earth). When God created the man and the woman in the divine image, it meant that they were created for relationships.

Both the Priestly (Gn 1:1-2:3) and the Yahwist (Gn 2:4b-25) accounts describe *relational* aspect of the image of God. As an aspect of this relationality, God can communicate with men and women. In Genesis 1:29-30, God speaks directly to them—communication represents a significant form of relating. Because human beings are made in God's image, we can enter into a relationship with God, and in fact, this relationship with God is the

[1]Barth, *Church Dogmatics* II/1, *The Doctrine of God*, 86ff, 129ff, and his "Nein!" to Emil Brunner, "No! Answer to Emil Brunner," *Karl Barth: Theologian of Freedom* (ed. Clifford Green; London: Collins, 1989), 151-67; on Vatican I, see Pannenberg, *Systematic Theology* 1:75-76.

highest call of human beings. Jesus implies this relationship with an invitation: "*Love* one another as I have loved you" (Jn 15.12, italics added) and even goes so far as to call his followers "friends" (Jn 15.15), indicating how intimate this relationship can be.

In Genesis 1:27, creation as male and female implies that human beings are to relate to one another, for which marriage (Gn 2:24) is the most definitive human institution. Relationality is also demonstrated by the fact that both male and female are created in God's image, and therefore neither is definitively *the* human being. Stated positively, humanity is only adequately represented by both sexes.

In Genesis 2, Adam is told to have "dominion" over the animals—better understood as stewardship like a good king—and names the animals (Gn 2:19).[1] In the subsequent curses in Genesis 3 after Adam and Eve have eaten the fruit, they will experience a disrupted relationship with earth, i.e., that there will be toil in farming, and the ground "will produce thorn and thistles" (Gn 3:17-19 NIV).

---

[1] The scholarship on this passage is immense so I will simply quote Douglas John Hall, "Under the conditions of imperial Christianity, it was not stewardship but lordliness that appealed to the mentality of the church's policy makers. Thus, historic Christianity has seemed either to ignore and escape from the world, or else wish to possess it." Douglas John Hall, *The Steward: A Biblical Symbol Come of Age* (rev. ed; Eugene: Wipf & Stock, 2004), 82. Here, I am reminded of René Descartes' notorious phrase that we are "masters and possessors of nature." Hall continues his analysis and reclaiming of the concept of human stewardship: "it means that we must take in action role in tending creation and abandon forms of religion that denigrate the natural world, that view the world as primarily a cache of resources to be exploited for human ends" (Hall, *The Steward*, 82).

One of the core texts of the Torah, the Ten Commandments (Ex 20 and Dt 5) exemplifies these four relationships established in creation (with God, with other human beings, with ourselves, and with the rest of creation). The first four commandments begin with God: "I am the Lord your God, who brought you out of the land of Egypt, out of the house of slavery; you shall have no other gods before me" (Ex 20:2-3). As creatures, our essential relationship is with our Creator. As distinct from other creatures, we can return praise to God. The "second tablet" of the Decalogue addresses human relationships such as prohibitions against stealing and committing adultery. And there is a hint of the relationship with the rest of creation: the Sabbath command sets up not only rest for human beings but also for the "ox or your donkey, or any of your livestock" (Dt 5:14). There is also an implied relationship with self that is necessary for human moral reflection, which comes most clearly into view in Paul's tortured self-reflection in Romans 7:7-25[1] that encapsulated his cry of individual incomprehension: "I do not understand my own actions" (Rom 7:15). Proper relationality means harmony while disruption, disharmony, and incomprehension reveal a tortured and sinful relation. But for the purposes here, it is a relation nonetheless.

This essentially human relationality—especially in our relation to God—sets up a natural knowledge of God. Romans 1:18-20 and 2:14-15 constitute the *locus classicus* for the natural knowledge of God, or indeed, a natural theology.[2] In Romans 1:19-20, Paul notes this

---

[1] Whether Paul is speaking pre- or post-conversion is not relevant in this exposition.

[2] Pannenberg, *Systematic Theology* 1: 95ff.

awareness: "For what can be known about God is plain to them, because God has shown it to them. Ever since the creation of the world his eternal power and divine nature, invisible though they are, have been understood and seen through the things he has made." As Paul lays out his case for why all stand before God in need of Christ's redemption, he argues that all people know "God" (or perhaps better "god"—the garden variety word *theos* is used here). As James D. G. Dunn writes, "some sort of natural theology is involved here.... Paul is certainly conversant with and indeed indebted to a strong strand of like-minded Hellenistic Jewish wisdom theology." Dunn notes Wisdom of Solomon 12-15, especially verses 19-32.[1] Still, for Paul, this knowledge remains relatively vaguely—only his "eternal power" and "deity" or "divine nature."

Additionally, in Romans 2, Paul is arguing that both Jews and Gentiles stand universally in need of Christ's redemption. He is moving toward the key statement, "All have sinned and fall short of the glory of God" (Rom 3:23), to be resolved by the redemption in Christ. "But God proves his love for us that while we were still sinners Christ died for us" (Rom 5:8). In the

---

[1] James D. G. Dunn, *Romans 1-8* (Word Biblical Commentary 38A; Dallas: Word, 1988), 56. A comparison text is Ecclesiastes 3:11 which asserts that God has put "eternity into our hearts." Overall I agree with the exposition of Joseph Fitzmyer that Paul is indicating humankind has some innate knowledge of God. See Joseph Fitzmyer, *Romans: A New Translation and Commentary* (Anchor Bible; New York: Doubleday, 1992), 273-79. I cannot agree with Karl Barth, "Paul does not dream of paying the Gentiles anything resembling a compliment and of trying to find in their religions some point of contact for the understanding of the Gospel...." Karl Barth, *A Shorter Commentary on Romans* (trans. D. H. van Daalen; Richmond: John Knox, 1959), 29.

course of this argument, he appeals to the conscience of the Gentiles,[1] and their ability to do "what the law requires." Specifically, in Romans 2:14-16, Paul wrote:

> When Gentiles, who do not possess the law, do instinctively what the law requires, these, though not having the law, are a law to themselves. They show that what the law requires is written on their hearts, to which their own conscience also bears witness; and their conflicting thoughts will accuse or perhaps excuse them on the day when, according to my gospel, God, through Jesus Christ, will judge the secret thoughts of all.

Paul's point here is not about natural theology *per se,* but that the Gentiles have some innate or natural knowledge of God's moral will. As Dunn notes, Paul is referring to Stoic or Jewish Wisdom ideas, or both;[2] but in any event, he clearly presents a natural knowledge of God. He is expanding on the previous statement in 1:18-20 about God's "invisible power and deity" to include a sufficient knowledge of God's "law" or morality by which all will be judged.

Put together, these biblical passages assert that we are created for God, that we know the general nature of God (especially his power and otherness as deity), and that we have a moral conscience. These all inform my formulations of the *sensus divinitatis.*

### The Theological Tradition on the Natural Knowledge of God

---

[1] Here I follow Dunn (*Romans,* 1:100), these are not Gentile Christians; *contra* C. E. B. Cranfield, *The Epistle to the Romans* (ICC; Edinburgh: T&T Clark, 1975), 1:156.

[2] Dunn, *Romans 1-8,* 1:105-06.

The theme of the natural knowledge of God has made its way into subsequent Christian theology. As Augustine wrote early in the fifth century, in his beautiful opening prayer to *Confessions*, "Lord, you have made us for yourself, and our hearts are restless until they rest in you."[1] This is of course a prayer and therefore occurs within faith—it is not therefore technically a proof—and in it Augustine gives our natural yearning for God both an existential and creational caste.

More philosophically, the great thirteenth century Roman Catholic theologian, Thomas Aquinas (who thus wrote before the Protestant/Catholic divide), offered an outline of the famous "Five Ways" in his *Summa Theologica* (which are more fully developed in *Summa contra Gentiles*). First, there is the *Argument from Motion:* since everything that moves is moved by another, there must thereby exist an Unmoved Mover. Second, the *Argument from Efficient Cause*: the sequence of causes that make up this universe must have a First Cause. Third, the *Argument to Necessary Being:* since all things that exist are dependent on other things for their existence, there must exist at least one thing that is not dependent. This then is a Necessary Being. Fourth, the *Argument from Gradation:* since all things that exist can be compared to such qualities as degrees of goodness, there must exist something that is an Absolutely Good Being. Finally, the *Teleological Argument*: the intricate design and order of existent things and natural processes imply that a Great Designer exists. Whether or not these Five Ways are maligned or praised, they have offered a definitive

---

1 Augustine, *Confessions* (trans. R. S. Pine-Coffin; Harmondsworth: Penguin, 1961), 1.2.

outline for subsequent thinkers who make philosophical arguments for God's existence and for our natural knowledge of God. In fact, they really constitute a summary of what would have been known to his students reading the *Summa* and therefore are not a full-blown proof. Thomas bases the Five Ways on the conviction that human beings have knowledge *that* God exists, although revelation is needed to know who God is.[1]

In the seventeenth century—right at the flowering of modern science—the mathematician Blaise Pascal offered another proof for God. He began, in a similar vein to Augustine with our existential search for rest: "By nature, we all seek happiness." But where do we seek it? "Some seek the good in authority, some in intellectual inquiry and knowledge, some in pleasure." Pascal continued by observing that all these various potential sources for happiness, for a beautiful life, leave us craving for more. He pondered what that meant:

> What else does this craving, proclaim but that there was once in man a true happiness, of which all that now remains is the empty print and trace? This he tries in vain to fill with everything around him... since this infinite abyss can be filled only with an infinite and immutable object; in other words by God himself.[2]

C. S. Lewis echoed this conclusion about three hundred years after Pascal with a simple, logically compelling phrase: "If I find in myself desires which nothing in this world can satisfy, the only logical

---

[1] *Summa* 1.2.3.

[2] Blaise Pascal, *Pensées* (trans. A. J. Krailsheimer; Harmondsworth: Penguin, 1966), 45.

explanation is that I was made for another world."[1] Lewis believed that this argument from desire constitutes one of the strongest proofs for God's existence.[2]

It may surprise some in the Reformed tradition—at least those who have read Karl Barth's cavils against "natural theology"—that the seminal voice of Reformed theology, John Calvin, wrote similarly of the "awareness of divinity." Calvin was not out to prove God, but to state that inherent in human existence is a basic, vague, and powerful natural knowledge of God. Indeed, in Calvin's vastly influential 1559 *Institutes of the Christian Religion*, he wrote, "There is within the human mind, and indeed by natural instinct, an awareness of divinity."[3] This awareness of divinity, or *sensus divinitatis,* is "beyond dispute" according to Calvin. It is this formulation of the natural knowledge of God that I follow in this chapter.[4]

One final note: in my view, instead of working as *proofs* for God's existence, these arguments demonstrated that the natural knowledge of God *witnesses* to the God whom the church confesses to be Maker of heaven and earth. They may in fact work as proofs, but that is not

---

[1] C. S. Lewis, *Mere Christianity* (New York: MacMillan, 1960), 114.

[2] See my exposition of Lewis in *C. S. Lewis and the Crisis of a Christian* (Louisville: Westminster John Knox, 2014), 45-59.

[3] Calvin, *Institutes*, 1.3.1.

[4] This may be what the Roman Catholic theological giant Karl Rahner is after in his transcendental "openness to being," but I find his reflections too tinged with Kantian notions to be sufficiently biblical or (to me) convincing. See *Foundations of Christian Faith: An Introduction to the Idea of Christianity* (trans. William V. Dych; New York: Crossroads, 1982).

my focus; instead I am arguing that the doctrine of creation—that God created this world and us as part of it—implies that all human beings possess a natural knowledge of God. It is not that we see this natural knowledge, and therefore God exists. Instead, when we see the world as created by God, we realize that this *sensus divinitatis* exists in all people.

**Beauty in Science and Theology as a Witness to the Natural Knowledge of God**

For a scientifically informed systematic theology, one promising nexus for the *sensus divinitatis*, or the natural knowledge of God, is the perception of beauty.[1] God has created this world beautiful—as it reflects the divine source of beauty—and whether explicated as a theological category or not, that beauty shines through the natural world. And it is a beauty that scientists and believers both perceive.

Through creation, human beings experience beauty. As Gerald Manley Hopkins, the profound nineteenth century poet, intones: "Give beauty back, beauty, beauty, beauty, back to God, beauty's self and

---

[1] For this subsection, cf. Alister McGrath, *A Scientific Theology: Nature*, 1:232-40. Incidentally, I am not making a proof for God's existence from the existence of beauty although many excellent Christian thinkers have done so. Consider Augustine's argument, "The world itself, by its well-ordered changes and movements, and by the fair appearance of all visible things, bears a testimony of its own, both that it has been created, and also that it could not have been created save by God, whose greatness and beauty are unutterable and invisible." *City of God* XI.4, cited by Thomas C. Oden, *The Living God, Systematic Theology* (San Francisco: HarperSan-Francisco, 1987), 1:170.

beauty's giver."[1] The Psalmist declares God's beauty: "One thing I asked of the Lord, that will I seek after: to live in the house of the Lord all the days of my life, to behold the beauty of the Lord, and to inquire in his temple" (Ps 27:4).

What is beauty? According to the ancient tradition, beauty is transcendental—like goodness, unity, and truth. Though it is ultimately indefinable (because it cannot be set within a larger category), Plato offers three markers for beauty: order, symmetry, and proportion; similarly, Thomas Aquinas highlighted integrity, consonance, and clarity (*integritas, harmonia, claritas*).[2] Thomas Oden offers something closer to a definition: "Beauty is that quality or combination of qualities within a thing that gives pleasure to the senses or pleasurably exalts the mind of spirit."[3]

Beauty arises for both theologians and scientists through rightly grasping and theorizing about their objects of study. Beauty thus leads to truth, and beauty provides a lure for study. In this sense, it is *telic*, that is, leading human beings toward a preferred future. For theologians, it means grasping God's true nature, God's creation, and our ethical life. For scientists, it is rightly perceiving, and theorizing about, nature. When this perception is made there is discovery, which is accompanied by a sense of completeness. In these and

---

[1] Gerard Manley Hopkins, "The Golden Echo," *Selected Poems of Gerard Manley Hopkins* (ed. Bob Blaisdell; Mineola: Dover, 2011).

[2] See, for example, *The Cambridge Dictionary of Philosophy* (2nd edition; ed. Robert Audi; Cambridge: Cambridge University Press, 1999), 76.

[3] Oden, *The Living God*, 168.

other ways, beauty represents a common value for scientists and theologians.[1]

One of the most important, and underappreciated, voices on the importance of beauty for theology is Jonathan Edwards.[2] Beauty captivated this eighteenth century theologian and philosopher—the beauty of the natural world, of God, and of life lived to God's glory. Edwards spoke of a particular early experience where contemplation led him "into a kind of vision... of being alone in the mountains, or some solitary wilderness, far from all mankind, sweetly conversing with Christ, and wrapped and swallowed up in God."[3] Steeped in the observation of nature that marked the exuberant scientific explosion following Newton's impressive discoveries and seminal theories, Edwards gloried in the beauty of nature. It is worth noting Puritan pastors, as some of the most educated members of their day, regularly found numerous causes for reflection on God, nature, and their relationship through natural philosophy. They quite naturally engaged in what today we call "theology and science."

---

[1] These themes are echoed in the substantial Catholic voice of Hans Urs von Balthasar, specifically through his magisterial *The Glory of the Lord: A Theological Aesthetics* (7 vols.; Edinburgh: T & T Clark, 1982-91). I cannot overestimate Balthasar's contributions to a theological aesthetics, and my debt to his theological aesthetics is substantial.

[2] For example, in Balthasar's seven volumes I cannot find a single line on Edwards. For the importance of beauty in his theology, see Louis J. Mitchell, *Jonathan Edwards and the Experience of Beauty* (Studies in Reformed Theology and History 9; Princeton: Princeton Theological Seminary Press, 2003).

[3] George Marsden, *Jonathan Edwards: A Life* (New Haven: Yale University, 2007), 42.

Edwards's *natural beauty* "consists of a very complicated harmony; and all the motions and tendencies and figures of bodies in the universe are done according to *proportion*, and therein lies their beauty."[1] (The echoes of the classical tradition of beauty as *proportio* are unmistakable.) He also underscored the importance of God's work as Creator of this cosmos:

> For as God is infinitely the greatest being, so he is allowed to be infinitely the most beautiful and excellent: and all the beauty to be found throughout the whole creation is but the reflection of the diffused beams of that Being who hath an infinite fullness of brightness and glory.[2]

In his philosophical-theological writings, Edwards maintained a lifelong "preoccupation with beauty, excellence, and the goodness of creation."[3] Finding beauty is at the core of his definition of the spiritual life. To be fully alive as a human being is to be drawn into beauty. Beauty in nature evokes a deeper praise for the Source of beauty. In this way, creation leads back to the Creator.[4]

---

1 "The Mind," 335; cited in Mitchell, *Jonathan Edwards*, 4, emphasis mine.

2 John E. Smith, Harry S. Stout, and Kenneth P. Minkem, eds., *A Jonathan Edwards Reader* (New Haven: Yale University Press, 2003), 252.

3 Smith, *Edwards Reader*, xii.

4 For example, Edwards also linked the beauty of God and the beauty of creation with the beauty of our ethical life, a theme worth developing separately, although not sufficiently related to the topic at hand. Another more recent contribution comes from the contemporary English theologian and philosopher, Keith Ward, quite simply speaks of God as "absolute beauty and goodness" in *The Big Questions in Science and Religion* (West Conshohocken: Templeton, 2008), 192.

What do natural scientists say? Remarkably, in reading some scientists' descriptions of their own work, I have discovered a remarkable similarity with theology, such that I could transpose words between theology and science and the statements would sound nearly identical. The beauty of scientific work is to understand nature rightly and the way it fits together. This common value provides a stimulating locus for collaboration of theology and science. Beauty lures us to truth—both in that its innate pleasure motivates human beings to discover truth and that beauty and truth conform to one another. As Aquinas wrote, "The supreme beauty of human nature consists in the splendor of knowledge."[1] Beauty is critical to all human knowledge, including the natural sciences. Richard Feynman once wrote, "You can recognize truth by its beauty and simplicity."[2] Beauty, as both scientists and theologians know, leads to truth.

The Nobel laureate, Subrahmanyan Chandrasekhar, presented an important study in "Beauty and the Quest for Beauty in Science" by pursuing "the extent to which the quest for beauty is an aim in the pursuit of science."[3] For example, Henri Poincaré, when answering the question of why scientists study nature at all and how they select facts when formulating scientific theory,

---

[1] *Maxima pulchritude humanae consistit in splendore scientiae.* Thomas Aquinas, *On Evil* (trans. Richard Regan; ed. Brian Davies; Oxford: Oxford University Press, 2003), 4.2 obj. 17.

[2] Cited in Robert Augros and George Stanciu, *The New Story of Science* (Washington, DC: Regnery, 1984), 39.

[3] This represents his seminal 1979 lecture at the Fermi National Accelerator Laboratory. In *Truth and Beauty: Aesthetic and Motivations in Science* (Chicago: University of Chicago Press, 1987), 59. See also, J. W. McAllister, "Truth and Beauty in Scientific Reason," *Synthese* 78.1 (1989): 25-35.

noted: "The scientist does not study nature because it is useful to do so." He continued and thereby countered a purely instrumentalist approach to scientific work and simultaneously described the way that beauty motivates scientific discovery and offers scientists what I call a "*telos*"[1] or motivation:

> He studies it because he takes pleasure in it; and he takes pleasure in it because it is beautiful. If nature were not beautiful, it would not be worth knowing and life would not be worth living.... I mean the intimate beauty which comes from the harmonious order of its parts and which a pure intelligence can grasp.[2]

Poincaré points to harmony or consonance as a central feature of beauty. Beauty also implies pleasure (which has constituted key elements of theories of beauty for centuries), and thus scientists realize the pleasure of their work in the realization of harmony. This beauty sustains scientists' research even in spite of the rigors of their work: "Intellectual beauty," he continued, "is self-sufficing, and it is for it, more perhaps than for the future good of humanity, that the scientist condemns himself to long and painful labors."[3] Similarly, Werner Heisenberg wrote about the connection between discovering the nature of quantum reality and its beauty. Beauty for Heisenberg is surprising and objective. As he describes it, he did not

---

[1] Greg Cootsona, "How Nature and Beauty Can Bring Scientists and Theologians Together," *Theology and Science* 9.4 (2011): 379-93, esp. 381, 384.

[2] Henri Poincaré, *Science and Method* (New York: Dover, 2003), 22.

[3] Poincaré, *Science and Method*, 22.

impose beauty, but *discovered* this beauty in the midst of looking at energy at the quantum level:

> I had the feeling that, through the surface of atomic phenomena, I was looking at a strangely beautiful interior, and felt almost giddy at the thought that I now had to probe this wealth of mathematical structure nature had so generously spread out before me.[1]

This pursuit and discovery of beauty has certainly motivated key scientists. I could multiply quotes, but will simply note Einstein's use of beauty in formulating both the special and general theories of relativity. Helen Dukas and Banesh Hoffmann summarized Einstein's work: "The essence of Einstein's profundity lay in his simplicity; and the essence of his science lay in his artistry—his phenomenal sense of beauty."[2] It was that sense of beauty that led him to reformulate our understanding of the cosmos. The particular motivation of beauty for scientists, as Poincaré describes it, i.e., in grasping the harmonious order of the cosmos. Indeed, in *Adventures of Ideas,* Alfred North Whitehead pointed to this ordering function of scientific and artistic pursuits. As he wrote, "Science and art are the consciously determined pursuit of Truth and of Beauty."[3] It is beauty that lures us and that makes truth worth discovering.[1]

---

[1] Werner Heisenberg, *Physics and Beyond: Encounters and Conversations* (New York: Harper & Row, 1971), 68, cited in McGrath, *Scientific Theology,* 1:239.

[2] Helen Dukas and Banesh Hoffmann, *Albert Einstein: Creator and Rebel* (New York: Penguin, 1972), 3.

[3] Whitehead, *Adventures of Ideas,* 272. Perhaps Whitehead did not know, or care to know, that a significant component of twentieth century art and aesthetics has abandoned the search for beauty. See Stolnitz, Jerome, "Beauty," *The Encyclopedia of Philosophy* (New York:

We are coming to a point where it becomes less fruitful to speak of separate directions for theology and science, but in fact, the locus of common understanding and more importantly, motivation. The noted physicist George Ellis has presented beauty as the highest level of human knowledge: "I believe that for many the experience of great beauty is an immediate striking way of experiencing transcendence."[2] Ellis noted that this leads many people to "genuinely spiritual experience."[3] In Ellis (and to some degree in Whitehead), I see the confluence of these disciplines, science and theology, in one person.

## The Cognitive Science of Religion and Calvin's *Sensus Divinitatis*

---

MacMillan and Free Press, 1967), 1: 266. Whitehead then connected this ordering with God's work with the world. At the heart of Whitehead's understanding of the God-world relation, he concluded that God "does not create the world, he saves it: or, more accurately, he is the poet of the world, with tender patience leading it by his vision of truth, beauty, and goodness." See Alfred North Whitehead, *Process and Reality* (Corrected ed.; ed. David Ray Griffin and Donald W. Sherburne; New York: Free Press, 1978), 526 (346).

[1] See Alejandro García-Rivera's reflections on the importance of the beauty as that which moves "the heart," or the center of human action in *The Community of the Beautiful: A Theological Aesthetics* (Collegeville: Liturgical Press, 1999), 158ff.

[2] "Faith, Hope, and Doubt in Times of Uncertainty." This was a lecture presented to the Australia Yearly Meeting of the Religious Society of Friends (Quakers), in which he spoke of his study of nature as a scientist and the way it ultimately led to discovering beauty. See the complete published remarks in George Ellis, *Faith Hope and Doubt in Times of Uncertainty: Combining the Realms of Scientific and Spiritual Inquiry* (Queensland, Australia: Interactive Publications, 2008).

[3] "Faith, Hope, and Doubt."

An evolutionary understanding of the development of the human brain provides another, and more specific, starting point for a scientifically based natural knowledge of God. Justin Barrett, through his work in developing a Cognitive Science of Religion, uses the findings of the cognitive sciences to argue that evolution has developed human beings so that we implicitly see purposes in events, or are predisposed toward teleology. "Evidence exists that people are prone to see the world as purposeful and intentionally ordered,"[1] which naturally leads to belief in a Creator. For example, preschoolers "are inclined to see the world as purposefully designed *and* tend to see an intelligent, intentional agent behind this natural design."[2] Some use this tendency to impugn belief in God—i.e., we cannot help but believe. Instead I am arguing here that it is part of God's creation. We are created with an openness to belief. Another area of research suggests that evolutionary pressures, particularly the human need toward cooperation as it leads to survival, produces a common stock of morality; "a recurring theme is that humans seem to naturally converge upon a common set of intuitions that structure moral thought," such as "it is wrong to harm a nonconsenting member of one's group."[3] Andrew Newberg and Eugene D'Aquili have also studied brain activity during meditation and prayer

---

[1] Barrett, *Cognitive Science*, 59.

[2] Barrett, *Cognitive Science*, 71, and *Born Believers: The Science of Children's Religious Belief* (New York: Free Press, 2012). This feature of early childhood has been termed "promiscuous teleology" by the psychologist Deborah Kelemen (in Barrett, *Cognitive Science*, 70).

[3] Barrett, *Cognitive Science*, 86.

and found a remarkable cognitive function that supports belief in God.[1]

Certainly, there are similarities with John Calvin's *sensus divinitatis*, which points to a sense of the numinous, powerful and brooding. "Where can I go from Your presence? Where can I flee from Your spirit?" cries the psalmist in Psalm 139. It is the feeling of being out in a forest at night, knowing that no one is there, but feeling *something*. Often this experience can frighten us. And yet it also provides a witness to the natural knowledge of God. To be clear, God has used the process of evolution to implant this natural awareness.

**A Theological Critique of the *Sensus Divinitatis***

What is an appropriate theological appraisal of Calvin's *sensus divinitatis*? How does Calvin himself understand this sense of the divine? What critique does he offer? And what is the proper place of the *sensus divinitatis* for a scientifically informed ecclesial theology? Calvin continued his reflections on the *sensus divinitatis* by offering some caveats:

> Though the conviction may occasionally seem to vanish for a moment, it immediately returns, and rushes in with a new impetuosity, so that any interval of relief from the gnawings of conscience is not unlike the slumber of the intoxicated or the insane, who have no quiet rest in sleep, but are continually haunted with dire horrific dreams. Even the godless themselves,

---

[1] Andrew Newberg and Eugene D'Aquili, *Why God Won't Go Away: Brain Science and the Biology of Belief* (New York: Ballantine, 2002). See also Newberg, *Principles of Neurotheology* (Ashgate Religion and Science Series; Burlington: Ashgate, 2010).

therefore, are an example of the fact that *some idea of God* always exists in every human mind.[1]

The phrase "some idea of God" is instructive—Calvin emphasizes that the sense of the divine is ephemeral and elusive; he also writes that this *sensus divinitatis* is "fleeting and vain."[2] This is certainly not a sturdy foundation for faith. It is the general awareness of a Supreme Being, God's "eternal power and deity" which Paul describes in Romans 1. Though universal and powerful, this general sense of God has a remarkable malleability.

Along with Michael Welker,[3] I argue that this sense of the divine, however, remains powerful but problematic. Welker cites Job 19:6, 8: God "closed his net around me.... He has walled up my way so that I cannot pass, and he has set darkness upon my paths." This vague sense of deity can even terrify. As Calvin writes:

> The most audacious despiser of God is most easily disturbed, trembling at the sound of a falling leaf. How so, unless in vindication of the divine majesty, which smites their consciences the more strongly the more they endeavor to flee from it. They all, indeed, look out for hiding-places where they may conceal themselves from the presence of the Lord, and again efface it from their mind; but after all their efforts they remain caught within the net. Though the conviction may occasionally seem to vanish for a moment, it immediately returns, and rushes in with new impetuosity, so that any interval of relief from the gnawing of

---

[1] Calvin, *Institutes* 1.3.2, italics added.

[2] Calvin, *Institutes* 1.3.3.

[3] Michael Welker, *Creation and Reality* (trans. John F. Hoffmeyer; Minneapolis: Fortress, 1999), 23-28.

> conscience is not unlike the slumber of the intoxicated or the insane, who have no quiet rest in sleep, but are continually haunted with dire horrific dreams. Even the wicked themselves, therefore, are an example of the fact that some idea of God always exists in every human mind.[1]

From this vague concept ("some idea of God") human beings can never distinguish between fears and fantasies and true knowledge. They may continue to develop a neurotic piety: "Those therefore, who set up a fictitious worship, merely worship and adore their own delirious fancies"—a piety that leads into idolatry—"indeed, they would never dare so to trifle with God, had they not previously fashioned him after their own childish conceits."[2] And later, "Even idolatry is ample evidence of this conception."[3]

Calvin's language is characteristically strong and largely negative (Calvin could never be accused of an inflated view of human nature). Nevertheless, building a religious, or more contemporarily, "spiritual" practice from the *sensus divinitatis* has many of the elements of idolatry in that it often leaves human beings exactly where they started. As Lewis pointed out in an address to the Oxford Socratic Society, this vague sense of the divine can be highly manipulated and is even dangerous, pliable to all sorts of distortions. It cannot ultimately convert us to the good. Lewis calls this a "minimal religion." It leaves both Nazis and altruists as they started, except now with a veneer of belief and an assurance that what they already do now has divine

---

[1] Calvin, *Institutes* 1.3.3.

[2] Calvin, *Institutes* 1.4.3.

[3] Calvin, *Institutes* 1.3.1.

endorsement. "The minimal religion will, in my opinion, leave us all doing what we were doing before."[1] We therefore need more clarity for informed, and ultimately beneficial belief. It can be the basis of nature-worship, built on a sense of numinous natural world. It can be a brash, hedonistic worship of self, embodied in the basest forms of New Age spirituality. Even the Nazis propagated an appreciation for what "God is doing through the German *Volk*" and supported it with the powerful, but vague feeling of the Numinous working to renew the German civilization. It can also be named "transcendence"[2] or channeled in a variety of ways.

I need to summarize: This *sensus divinitatis* opens us to belief in God. We find hints in cognitive science and the scientists' pursuit of beauty. Nonetheless, it is a vague awareness that can neither *prove* God, nor can it give us fully developed attributes of God. And the specific problems of the *sensus divinitatis* reveal the more general weakness of natural theology. Nature gives us both stunning sunsets and devastating hurricanes, fertile farmlands and wind-swept dustbowls, impressive mountain peaks and deadly volcanoes. Nature's supporting data present evidence of two incompatible visions: the gracious, loving God and an angry, evil deity. Pascal, who plumbed the depths of such natural proofs for God, grasped the essential weakness of this approach.

> I wonder at the hardihood with which such persons undertake to talk about God. In a treatise addressed to infidels they begin with a

---

[1] C. S. Lewis, *God in the Dock: Essays on Theology and Ethics* (ed. Walter Hooper; Grand Rapids: Eerdmans, 1970), 141.

[2] Here I am thinking of Richard Dawkins, *The God Delusion*.

> chapter proving the existence of God from the works of Nature... this only gives their readers grounds for thinking that the proofs of religion are very weak... It is a remarkable fact that no canonical writer has ever used Nature to prove God.[1]

This *sensus divinitatis,* though part of our creation, leaves us open for God. It also, however, leaves human beings with a desire for clarity.

### Coda
### A Christological Reconstruction

What then are the purposes of nature and this natural awareness of divinity in leading us to God? It is not a proof, but a *witness,* a support for the God revealed in Jesus Christ. Christian believers fill in a natural awareness of God with Christological specificity. Only after we have heard God's voice to us in Jesus Christ, *then* we are able to proclaim with the psalmist, "The heavens are proclaiming the glory of God" (Ps 19:1). This concept is ultimately what Ian Barbour has termed, not a "natural theology," but a "theology of nature."[2] We see the world through our belief in a good Creator.

---

[1] Cited in Lewis, *Problem of Pain* (New York: Macmillan, 1971), 242-43.

[2] "Instead of a natural theology, I advocate a theology of nature, which is based primarily on religious experience and the life of the religious community but which includes some reformulation of traditional doctrines in the light of science. Theological doctrines start as human interpretations of individual and communal experience and are therefore subject to revision. Our understanding of God's relation to nature always reflects our view of nature." Ian Barbour, *Religion and Science: Historical and Contemporary Issues* (rev. ed.; San Francisco: HarperOne, 1997), 247.

Scripture, as Calvin concluded, becomes the "spectacles" by which we view the world.[1]

Science acts in some ways, in describing this *sensus divinitatis*, to offer general revelation. Through general revelation, we can certainly find out truths about God, but those truths receive clarity through God's special revelation in history, especially depicted in the pages of the Bible. For example, we can find the beauty of God's design in the natural world through scientific work—and thus be led to conclude that God is an incomparable Designer. We can, however, only know that God's creation is Trinitarian through special revelation.

In my view, theology can and must journey beyond the strict domains of science, but that it must not contradict those findings, I conclude that we need Jesus to save the *sensus divinitatis*, because, as the church confesses, Jesus definitively reveals God. In this sense, Jesus Christ saves natural knowledge of God from its vagueness. Christ displays that there is ultimately no hidden God, as he is "the image of the invisible God" (Col 1:15). And so, with Barth I conclude, "The meaning of deity "cannot be gathered from any notion of supreme, absolute non-worldly being. It can be learned only from what took place in Christ."[2] Our natural knowledge of God needs to be clarified by Christ.

---

[1] Calvin, *Institutes* 1.6.1. Welker also builds from the analysis of Calvin to his reflections on the need for revelation; see *Creation and Reality*, 28ff.

[2] Karl Barth, *Church Dogmatics IV/1: The Doctrine of Reconciliation*, 177. One caveat here, following the apophatic tradition (cf. for example, Vladimir Lossky, *The Mystical Theology of the Eastern Church* [New York: St. Vladimir's Seminary,1997]), there is always more to God than can be revealed in one human life, even the life of

Thus, Christ saves a natural knowledge of God from vagueness and potentially pernicious misuse. After developing his highly elaborated Christocentric theology in Romans, Paul moves to the hortatory. He calls the Roman churches to be transformed or *meta-morphicized* (to transliterate the Greek) by urging them to take the *form* of Christ, who is also the goal of human yearning:[1] "Do not be conformed to this world, but be *transformed* by the renewing of your minds, so that you may discern what is the will of God—what is good and acceptable and perfect" (Rom 12:1-2, italics mine). Consequently, through Christ, the Church can be formed—as a community in worship and discipleship—from a vague, amorphous *sensus divinitatis,* into bearing the image of God to the world.

---

Jesus of Nazareth. Even the Gospel of John admits that "there were also many other things which Jesus did" (Jn 21:25).

[1] Cf. García-Rivera, who specifically ties beauty with being "formed" in *The Community of the Beautiful,* 175-80.

## *Panentheism*

## *Hindrance or Help?*

*John Culp*

Although panentheism has a long history,[1] it has attracted attention recently as a way to understand the nature of God's involvement in the world especially in the current science and religion discussion. Panentheism as a specific way of understanding God's involvement in the world has most frequently developed in Christian cultures, but manifestations of similar approaches can be found in many religions.[2] This examination of the potential of panentheism to facilitate discussion between religion and science will focus on the Christian tradition, although there may be significant insights to be gained from considering other religious traditions.

The current religion and science discussion manifests a variety of understandings of the nature of the relation between science and religion. Although the popular discussion often appears to demonstrate a conflict between religion and science or occasionally an

---

[1] For an older history, see Charles Hartshorne and William L. Reese, *Philosophers Speak of God* (Chicago: University of Chicago Press, 1953). For a more recent history, see John W. Cooper, *Panentheism: The Other God of the Philosophers: From Plato to the Present* (Grand Rapids: Baker Academic, 2006).

[2] Loriliai Biernacki and Philip Clayton, eds., *Panentheism across the World's Traditions* (Oxford: Oxford University Press, 2014) provides a survey of the presence of panentheistic understandings in a variety of religious cultures.

indifference to the other discipline, more constructive relationships do take place. Some common ground is needed in order for these more constructive relationships to develop more fully. Perhaps the most significant commonality would be a metaphysical common ground, a shared understanding of the basic structures of reality. Advocates of panentheism propose their understanding of the nature of God's relationship to the world as a way to respond to foundational issues in the religion and science discussion.

But a variety of metaphysical issues challenge the development of a common ground for science and religion. These issues include whether the nature of reality is causally closed or open to special actions by God, whether regularity or indeterminacy characterize reality at the basic levels, and whether reality is composed of unrelated substances or interrelated actualities. Epistemological issues related to different ways of knowing such as observation, logical relationships, and coherence also challenge any attempt to formulate a common ground for discussions between science and religion. Issues such as these challenge Christian thinkers to rethink the ways that they have understood and described the God/world relation. Panentheism may offer resources that would enable Christian thought to develop Christian theological understandings of the God/world relation in a culture that rejects supernaturalism, emphasizes divine immanence, and develops noninterventionist understandings of miracles. This paper examines various Christian concerns about panentheism, describes panentheistic responses to those concerns, and suggests some conclusions about the helpfulness of panentheism for Christian responses to our contemporary intellectual and cultural context.

But first, some preliminaries. One initial problem concerns the type of approach to evaluating panentheism in responding to the contemporary situation. The question of whether or not panentheism can be a resource for Christian thought provides a more constructive starting point than asking if panentheism is orthodox or not. To some extent trying to determine the orthodoxy of panentheism confronts the problem of what is meant by both "Christian" and "panentheism." Furthermore, starting with the possibility that panentheism is not orthodox may lead to overlooking resources that panentheism offers to the religion/science discussion.

Another important preliminary consideration is the variety of meanings of "Christian" and "panentheism" which require definition for the sake of clarity. "Christian" can involve adjectives such as classical, traditional, and evangelical. "Classical" Christian thought refers to the appropriation of philosophical understandings to describe God as omnipotent, simple, and similar abstract terms. "Traditional" or "Ecumenical" Christian thought is often understood as based on the ecumenical creeds. And "Evangelical" Christian thought emphasizes conversion, the Gospel, the Bible, and Christ's sacrifice.

"Panentheism" literally means "all in God." Thomas Jay Oord's identification of thirteen different understandings of the term "in" demonstrates the impossibility of understanding "in" simplistically.[1]

---

[1] Tom Oord's list of 13 meanings of "in" appears in Philip Clayton, "Panentheism Today: A Constructive Systematic Evalu-

Furthermore, different types of panentheism have been identified. Niels Henrik Gregerson describes three types of panentheism: soteriological, revelational, and dipolar. Soteriological panentheism understands the world's being in God as a gift and only what is redeemed by God's grace exists in God. Revelational panentheism finds that the Divine Spirit expresses itself in the world by going out of God and returning to God enriched by the experiences of world history. Dipolar panentheism, associated with Alfred North Whitehead, identifies God as having an eternal nature and a nature that preserves the world.[1] Owen Thomas adds eschatological panentheism as holding that all will be in God in the final fulfillment of the world in God.[2]

As important as distinctions are for both Christians and panentheists, a general meaning for these two positions will facilitate evaluating the helpfulness of panentheism in the religion and science discussion. In this essay, "Christian" refers to the broad Christian tradition rather than any of the specific expressions of the Christian tradition such as classical or Evangelical. Likewise, "panentheism" understands the relationship between God and the world in ways that find the world to be significant for God's existence as well as God being significant for the world's existence.

---

ation," *In Whom We Live and Move and Have Our Being: Panentheistic Reflections on God's Presence in a Scientific World* (ed. Philip Clayton and Arthur Peacocke; Grand Rapids: Eerdmans, 2004), 249-64, here 253.

[1] Niels Henrik Gregersen, "Three Varieties of Panentheism," *In Whom We Live and Move and Have Our Being*, 19-35.

[2] Owen C. Thomas, "Problems in Panentheism," *The Oxford Handbook of Religion and Science* (ed. Philip Clayton and Zachary Simpson; Oxford: Oxford University Press, 2006), 652-64, here 658.

**Concerns about Panentheism**

Questions about the possibility of utilizing panentheism in Christian thought began with the growing popularity of process theology in the 1950s and 1960s. Many critiqued the adequacy of process theology as a resource for Christian theology. These criticisms often applied to panentheism because critics and process theologians alike agreed that process theology was panentheistic in its understanding of the God/world relation. One of the most careful and thorough early expressions of concern about the panentheism of process theology came from Ronald Nash in several publications.[1] A pervasive concern for Nash was God's identity as Ultimate. Nash says that neither omnipotence nor omniscience essentially characterizes the panentheistic concept of God. God directs and cooperates with the world rather than controlling the world.[2] Further, God's relation to time compromises God's sovereignty because God is limited by time. God cannot know the future from a timeless eternity because God does not exist independently from the temporal events of the world.[3] The limited nature of God's power and sovereignty is further demonstrated in that God cannot

---

[1] For a summary of criticisms of Panentheism by other traditional Christians, see John Culp, "Panentheism," *The Stanford Encyclopedia of Philosophy* (Summer 2015 ed.), Edward N. Zalta (ed.), http://plato.stanford.edu/archives/sum2015/entries/panentheism/

[2] Ronald H. Nash, *The Concept of God* (Grand Rapids: Zondervan, 1983), 24, 113-14.

[3] Ronald H. Nash, "Process Theology and Classical Theism," *Process Theology* (ed. Ronald H. Nash; Grand Rapids: Baker, 1987), 3-29, here 11.

completely destroy evil.[1] Events and decisions in the world limit God's actions both because God does not act independently of the events of the world and because God depends upon events in the world to fulfill God's purposes. To speak of the world as God's body loses the distinction between God and the world.[2] For Nash, an adequate concept of God required an unlimited form of absolute attributes such as power and knowledge.[3]

In 2006, John Cooper published a comprehensive description of panentheism. For Cooper, the distinction between classical theism and panentheism results from their different sources of Plato and Plotinus. Following Plato, classical theism affirms an eternal transcendent reality. Panentheism, following Plotinus, accepts a chain of being where the One generates all reality including physical reality.[4] After describing panentheism, Cooper expressed concern about the adequacy of panentheism in relation specifically to classical Christian theism's assertions that God is transcendent, self-sufficient, eternal, and immutable.[5] The classical concept of God's transcendence does not ignore that God is immanent in the world. In this view, God is supernaturally present to all beings and events, empowering creatures. But in God's self, God is utterly transcendent, all-determining, and changeless.[6] Cooper acknowledges that panentheism makes an ontological distinction between God and the

---

[1] Nash, *Concept*, 29.

[2] Nash, *Concept*, 25.

[3] Nash, *Concept*, 165.

[4] Cooper, *Panentheism*, 15.

[5] John W. Cooper, "Panentheism: The Other 'God of the Philosophers': An Overview," *American Theological Inquiry* 1.1 (2008): 11.

[6] Cooper, "Panentheism: Overview," 12.

world and that God transcends the world. But for panentheism, the world is "in" God ontologically. This contrasts with classical Christian theism's unqualified distinction between God and the world, although they are intimately related.[1] Cooper maintains that God's transcendence infinitely exceeds God's immanence.[2] Panentheism fails to maintain the priority of divine transcendence because of the shared ontology of God and creation. The limited panentheistic understanding of divine transcendence leads to two practical problems. The first issue is the frequent rejection of miracles as supernatural due to understanding God's immanence as congruity with the natural order.[3] The second problem is that panentheism lacks the certainty of God's ultimate solution to the problem of evil. Cooper accepts that classical Christian theism's high view of divine power raises the problem of evil, but responds that classical Christian theism does offer a basis for hope because the final defeat of evil by God is possible.[4]

In recent articles and blogs, Roger Olson offers a carefully qualified developing response to panentheism. Olson affirms the traditional Christian understanding that holds that God transcends and does not depend upon the world. He understands panentheism to take a variety of forms, some of which are acceptable while others are not. God's dependence upon the world characterizes the inadequate forms of panentheism. These forms of panentheism believe that God cannot exist without generating a world and that God and the

---

[1] Cooper, *Panentheism*, 18.

[2] Cooper, *Panentheism*, 328.

[3] Cooper, *Panentheism*, 334.

[4] Cooper, *Panentheism*, 335.

world coexist and codetermine one another by metaphysical necessity.[1] Making the world necessary to God's being renders creation and salvation necessary rather than gracious.[2] If God *must* create and save the world, God is not free to *choose* to create or save.[3] Further, if salvation is not by divine grace,[4] God never intervenes in the world and much is lost of God's freedom and power.[5]

In a variety of ways, these concerns about the adequacy of panentheism as a way to understand God's relation to the world question the panentheistic understanding of God's transcendence. These concerns have both theological and metaphysical aspects. Nash's concern for transcendence in God's power and knowledge and Cooper's concern about a limited understanding of the difference between God and creation express theological concerns. Nash's concern for God's distinctive identity seeks to emphasize God's transcendence over the world. Olson finds that an inadequate concept of transcendence loses divine grace because creation and salvation are not the result of God's free choice. Further, many Christian theists find that divine transcendence is lost when panentheists accept

---

[1] Roger E. Olson, "A Postconservative Evangelical Response to Panentheism," *Evangelical Review of Theology* 85.4 (2013): 331.

[2] Roger E. Olson, "Does Love Ever Coerce? My Response to 'The Uncontrolling Love of God' by Thomas Jay Oord." http://www.patheos.com/blogs/rogereolson/ on (accessed Nov. 28, 2015).

[3] Olson, "Postconservative Evangelical Response," 335.

[4] Roger E. Olson, "What's Wrong with Panentheism?" http://www.patheos.com/blogs/rogereolson/2012/08/whats-wrong-with-panentheism/ (accessed Dec. 1, 2015.)

[5] Olson, "Postconservative Evangelical Response."

that God cannot completely defeat evil either by intervening in the world supernaturally or by ultimately ending all evil. Metaphysically, if the world-in-God is understood ontologically, the presence of the world as different from God implies a limit upon God's nature. More specifically, if the world is necessary as an intrinsic property of God, the necessity of the world limits God.[1]

Ultimately, however, the concern is not about rejecting or affirming divine transcendence. Rather, the concern lies in the nature of transcendence and how transcendence and immanence are related. For classical Christian theism, transcendence takes the basic form of separation from the world. While God is present in the world, God's presence results from God's transcendence. God's immanence is a consequence of God's transcendence. Cooper's concept of transcendence is that divine transcendence infinitely exceeds divine immanence, and Olson's understanding of transcendence rejects any divine dependence. Panentheism holds that divine transcendence and immanence are equally important rather than one being derived from the other. Although this equality of transcendence and immanence is described in different ways, God's relation to the world is equally important as God's distinction from the world. In some forms of panentheism, this equality of transcendence and immanence is described by God's dipolar nature. In other forms of panentheism, this equality is described as the necessity of the world for God's transcendence. In all

---

[1] Benedikt Paul Gocke, "Panentheism and Classical Theism," *Sophia* 52.1 (April 2013): 61-75 describes the panentheistic understanding of relation of the world to God as intrinsic and necessary.

the forms of panentheism, God both influences and is influenced by the world.[1] The basic concern of classical Christian theism is that panentheism, with its mutual relation of God and the world, fails to give priority to divine transcendence.

**Responses to Concerns**

Panentheists have responded to the theological concerns about divine transcendence in a variety of ways. If the concern about God's essential omnipotence and omniscience grows out of a concern about God's difference from the world, panentheists distinguish between God's identity and the identity of worldly realities in differentiating their position from pantheism. One difference between God and worldly realities is that God's existence is not limited temporally while worldly realities are temporally limited. In this distinction between God's unlimited existence and worldly realities, panentheists clearly affirm the infinity of God and finite nature of the world. Additionally, David Ray Griffin defends the distinction between God and the world by pointing out the numerical difference between God and the world, even if there is no ontological distinction.[2]

---

[1] For examples of these different ways of describing the balance of transcendence and immanence, see Gregerson, "Three Varieties of Panentheism," 22-23; Michael W. Brierley, "The Potential of Panentheism for Dialogue between Science and Religion," *Oxford Handbook of Religion and Science*, 639-40; and Wesley J. Wildman, "Mark Johnston's Naturalistic Account of God and Nature, Life and Death," *American Journal of Theology and Philosophy* 32 (2011): 186.

[2] David Ray Griffin, "Panentheism: A Postmodern Revelation," *In Whom We Live and Move and Have Our Being*, 36-47, here 44-45.

to any one time or time period. Value is not lost due to the passing of time. God not only preserves value that has been achieved; but in process panentheism, God overcomes evil by placing evil in contrast to good thus bringing good out of evil. Evil does not have the final word, but is overcome by serving as a contrast to the good, highlighting the good that has been actualized. However, this is not the destruction of evil nor a state where evil is no longer possible.

Panentheists have offered several responses to the metaphysical concern that the world limits God by being ontologically within God. Moltmann says that interpenetration of the world in God and God in the world preserves both the unity and difference between God and the world. God's essential nature is not changed by the world.[1] Instead, God's essential nature as love withdraws and enables the world to exist. Clayton affirms an ontological monism from which emerge new types of realities. These new realities do not control the reality from which they emerge. [2]

Clayton and Joseph Bracken further develop responses to the concern that panentheism's affirmation of the necessity of the world as an intrinsic property of God limits God. Clayton maintains that the involvement of the world in an internal relationship with God does not completely constitute God's being. God is both eternal and responsive to the world. The world constitutes God's relational aspect but not the totality of

---

[1] Jurgen Moltmann, *The Coming of God: Christian Eschatology* (trans. Margaret Kohl; Minneapolis: Fortress, 1996), 280, 295.

[2] Philip Clayton, "Conceptual Foundations of Emergence Theory," *The Re-Emergence of Emergence: The Emergentist Hypothesis from Science to Religion* (ed. Philip Clayton and Paul Davies; Oxford, Oxford University Press, 2006), 1-31, here 2-4.

God.[1] Bracken develops a systems understanding of the Trinity in which systems can interrelate without loss of transcendence. Systems, groups of realities, are more than the sum of their interrelated actual entities and are enduring ontological totalities. The Trinity as Father, Son, and Holy Spirit are not identical but are unified. Likewise, God and the world are unified without loss of identity.[2] Thus, the identities of both the world and God are preserved in their unity because God is more than the sum of the events in the world and endures even though the worldly realities do not endure.

**Concluding Evaluation**

Shifts in cultural context require rethinking concepts about how God relates to the world. Traditional doctrines such as incarnation, Trinity, and *theosis* all developed in order to understand God's relation to the world in differing contexts. Discussions today about miracles as intervention or nonintervention and about randomness and chance in relation to God's purposes express a changing context. Contemporary theological developments need to respond to the development of scientific thought in ways that recognize the regularities of nature while seeking to be faithful to prior understandings of divine action in the world.

The understanding of transcendence in early Christian thought responded to the philosophical context of Platonism. In seeking to articulate a concept of God

---

[1] Philip Clayton, "Panentheist Internalism: Living within the Presence of the Trinitarian God," *Dialog* 40.3 (Fall 2001): 210.

[2] Joseph A. Bracken, "Panentheism and the Classical God-World Relationship: A Systems-Oriented Approach," *American Journal of Theology & Philosophy* 36.3 (2015): 214, 224.

that would be meaningful in that context, it became important to describe God in ways that emphasized God's difference from—and separation from—the ordinary, daily existence of change and loss. Thus, all reality derived from the unchanging origin of a reality separate from the world.

This Platonic concept of transcendence differed from the concept of transcendence present in the ancient world of the Hebrew Bible and much of the New Testament world. The ancient world thought of existence as a struggle between various powers. This was most clearly expressed in various dualistic formulations such as good and evil. In that context, transcendence was expressed as God being the most powerful of all the powers.[1] Monotheism offered the most complete understanding of the power of God over all other powers. In responding to the new context of Platonic thought, the Christian tradition maintained the priority of God by affirming God's difference from the world more than God's ability to defeat minor deities and other powers. In responding to the contemporary context, which emphasizes both the regularity of the world and the importance of relationships, the significance of the reality of God must be articulated in a way that both acknowledges the importance of God for the world and God's relationship to that world.

Contemporary understandings of transcendence reflect a tension between transcendence as *relational* and transcendence as *separation*. The *Oxford English Dictionary* reflects this tension. The general definition of transcendence includes terms such as "surmounting,"

---

[1] Psalms 97:7 and 9 and Ephesians 1:20-21 provide examples of this perspective by stating that God is over all powers.

"rising above," and "surpassing," all of which imply relationships of comparison. Still, the definition of *transcendence* in relation to deity says, "The attribute of being above and independent of the universe."[1] Panentheistic concepts of reality and transcendence accept that reality and transcendence involve relationship. However, relations in transcendence can be either horizontal or vertical. Relational transcendence may be a limited type of transcendence if it is a horizontal transcendence among individuals. Human self-transcendence requires the experience of a relationship with another self in order to move beyond personal limits. Still, human transcendence requires a relationship with something other than human existence. Transcendence of nonultimate reality requires the experience of a relationship with Ultimate reality in order to even think of Ultimate reality. Likewise, Ultimate reality is not transcendent without an experience of nonultimate reality. There is no transcendence if there is only God. Thus the understanding of transcendence that says that divine transcendence is above and independent of the universe overlooks the relationship necessary for even vertical transcendence.

Panentheism acknowledges the importance of the relational nature of transcendence and responds to the contemporary context by balancing transcendence with immanence. This balance of transcendence and immanence avoids the difficulties with both the classical Christian understanding and certain contemporary understandings that are identified as postmodern,

---

[1] "transcendence, n." *Oxford English Dictionary Online.* (accessed 5/5/2016).

postsecular, or at times as naturalistic. The classical Christian tradition's emphasis upon divine transcendence defines God as an essentially an unrelated "other" in contrast to ordinary existence. God's existence does not require ordinary existence, although the world's existence depends upon God's existence. Divine immanence derives from God's transcendence. God is present to all of reality because God's transcendence enables God to relate to all of reality. But basing immanence on the separation of divine reality from ordinary reality results in an external relation between God and the world. The external nature of this relation limits God's presence because God is not affected by the world. The relation between God and the world is an asymmetrical relation in that God affects the world, but the world has no impact on God.

The postmodern theological tradition, in distinction from classical Christian theism, understands God as wholly immanent. This emphasis upon God's immanence appears in efforts to relate scientific work on evolution to theology by thinking of God's creative work as taking place within the world rather than from outside of the world. Creation through evolution rather than creation from nothing becomes the way of thinking about creation. Catherine Keller's concept of creation from chaos provides one example of this theological approach to evolution. Immanence is also the primary understanding of God in the "weak God" of postmodern thinkers such as Giles Deleuze, Gianteresio Vattimo, and John Caputo, who reject prioritizing divine transcendence when thinking about God. This emphasis upon immanence affirms the presence of God in the world while recognizing the need for a source of newness and novelty that is not limited by the past.

However, rather than an external reality providing novelty for the world, the basis for novelty is internal to reality. According to Daniel Barber, Deleuzian immanence, for example, affirms that God has a greater power because God can both generate and receive.[1] Immanence draws on the richness of creation rather than on a transcendent reality independent of creation. The particularity of Jesus exemplifies this richness of creation.[2] Similarly, Nancy Frankenberry calls for a new view of transcendence within immanence.[3] But, the emphasis upon immanence appears to limit the distinctiveness of God's role in the world to what was present in creation.

While panentheism offers a metaphysical basis for a balance between divine transcendence and immanence, panentheism does not provide a comprehensive theology. But its focus on the God/world relation does have implications for a broad range of Christian doctrines. For example, panentheists such as Moltmann and Bracken have utilized the concept of mutual relation in their treatments of the doctrine of the Trinity. Furthermore, some of those concerned about panentheism have expressed a tentative acceptance of panentheism in articulating a broad Christian theology. Cooper respects panentheistic theologies as authentically Christian provided that they are committed to Scripture

---

[1] Daniel Colucciello Barber, *Deleuze and the Naming of God: Post-Secularism and the Future of Immanence* (Edinburgh: Edinburgh University Press, 2014), 46-47.

[2] Barber, *Deleuze and the Naming of God,* 110, 118.

[3] Nancy Frankenberry, "Enduring Questions in Philosophy of Religion: A Response to Neville and Godlove," *American Journal of Theology & Philosophy* 37.1 (2016): 52.

and the ecumenical creeds.[1] Olson finds panentheism acceptable to evangelicals even if creation is inevitable as long as it maintains the paradox that inevitability does not require necessity.[2]

Panentheism's balance between immanence and transcendence has proven fruitful in relating religion and science. Arthur Peacocke, working as a scientist knowledgeable about theology, has utilized panentheism to relate religion and science by emphasizing divine immanence while retaining transcendence.[3] Peacocke rejects a mechanistic understanding of the world for the current understanding of the world as a unit of complex systems with a hierarchy of different levels that emerge from the complexity. This creative dynamic of the emergence of complexity is immanent in the world rather than external to the world.[4] God continuously creates from within the processes of the natural order rather than being external to the world. But God is not identified with the natural processes. God transcends the universe because God is infinitely more than the universe.

Panentheism may also prove helpful in understanding the relation between God's nature as

---

[1] Cooper, "Panentheism: Overview," 23-24.

[2] Olson, "A Postconservative Evangelical Response to Panentheism," 337.

[3] See Arthur Peacocke, *Creation and the World of Science* (Oxford: Clarendon Press, 1979); *Paths from Science towards God: The End of All Our Exploring* (Oxford: Oneworld, 2001); "Articulating God's Presence in and to the World Unveiled by the Sciences," *In Whom We Live and Move and Have Our Being*, 137–154; and "Emergence, Mind, and Divine Action: The Hierarchy of the Sciences in Relation to the Human Mind-Brain-Body," *Re-Emergence of Emergence*, 257–78.

[4] Peacocke, "Articulating God's Presence," 145-48.

related to the world and God's will as independent of the world. Questions about the relationship between God's nature and will have been implicit in the discussion about the availability of panentheism as a resource for Christian thought. Griffin alludes to this issue by identifying the world's causal pattern as belonging to the very nature of God. Any interruption of the world's causal pattern would violate God's nature.[1] Olson's emphasis upon divine freedom and his acceptance of Wolfhart Pannenberg's and Moltmann's emphasis upon God's nature as love as free also raises questions about the relation of God's nature to God's will.[2] God creates out of God's nature as love but does so freely rather than due to the nature of God's being.

Richard Kearney's *Anatheism* and Thomas Jay Oord's *The Uncontrolling Love of God* point to the importance of questions about the relation between God's nature and God's will in contemporary theology. Kearney describes God as a transcendence working in and through immanence.[3] For Kearney, God works through the acceptance of the diversity of religious traditions that opens up acceptance of the stranger and transcendence.[4] Oord bases both God's creating and continuing action in the world upon God's nature as uncontrolling love. For Oord, God is not free to love since God's nature is love. Loving is a necessary aspect

---

[1] David Ray Griffin, *Panentheism and Scientific Naturalism: Rethinking Evil, Morality, Religious Experience, Religious Pluralism, and the Academic Study of Religion* (Claremont: Process Century Press, 2014), 3.

[2] Olson, "A Postconservative Evangelical Response," 337.

[3] Richard Kearney, *Anatheism: Returning to God after God* (New York: Columbia University Press, 2011), 182.

[4] Kearney, *Anatheism*, 175.

of God's unchanging nature. God does choose how to love.[1] Both Kearney and Oord give priority to God's nature over God's will. In doing so, they make the issue of the relation between God's nature and will apparent. If God's nature as love results in the creation of a world and care for that world, then the relation between God and the world can no longer be understood as simply the result of a divine choice with no basis in God's nature. God as caring for the world then results in an internal relationship in which the world affects God because of who God is and not just because God has arbitrarily decided to relate to the world.

Theologically and practically the relation between God's transcendence and immanence derives from the relation of God's nature as love to God's loving action expressing God's will. God's immanence in creating and continuing the world expresses God's nature as lovingly creative. God as lovingly creative is affected by and responds to the creation. This nature of being affected by the world is recognized most fully through an ontologically based mutual relationship that maintains the distinct identities of God and the world. God's transcendence is manifested in God's acting freely within relationships without being restricted to those relationships. The reliability of God's nature as love cannot provide the basis for God's loving actions if those loving actions are caused by a will separate from God's nature. Those loving actions cannot be random and unrelated to a broader purpose. Thus, God's will to respond to specific situations requires direction by God's

---

[1] Thomas Jay Oord, *The Uncontrolling Love of God: An Open and Relational Account of Providence* (Downers Grove: IVP Academic, 2015), 161-62.

nature. And yet, simply granting priority to God's nature could lead to an isolation of God's nature from God's will and God's activity in the world. It appears that an ongoing interaction between God's nature and will is necessary. God's nature directs God's will, which in its actions expresses God's nature in new ways, thus influencing God's nature.

Panentheism's concept of a mutual relation between God and the world can aid Christian thought in speaking meaningfully to contemporary ways of thinking. The panentheistic mutual relation balances divine transcendence and immanence, preserving the claim that God is unique while embracing the intimacy of the relationship between God and the universe.[1] God and the world mutually influence each other. Panentheism does not result in a God who is unable to work in the world. But, God's action takes place through the world rather than from outside the world, and God suffers real loss when God's direction is rejected by the world. Panentheism retains past Christian affirmations of both God's priority for the world and God's activity in relation to the world, divine transcendence and immanence, in a balanced manner that avoids the problems of prioritizing either transcendence or immanence in the present context. Panentheism offers one way to support belief in the God who was longsuffering towards Israel and forgiving when Nineveh repented.

---

[1] Gregersen, "Three Varieties of Panentheism," 19.

## *A Process Thought Inquiry into Importance*
## *Religion as Constructive and Postmodern*

*John Becker*

The world in which we live and breathe has radically changed since the onset of the Enlightenment and the Industrial Revolution in Western Europe during the 17th century. In this radical shift, the ownership of defining "progress," rightly or wrongly, came under the sole tutelage of science and industry. One of the most disastrous results of this new ownership was the limited way "progress" was characterized, namely, as something quantifiable. After all, "numbers don't lie" as the expression goes. Numbers, according to this presupposition, are independent of the fallibility of humanity and, therefore, are often seen as an objective way to accurately measure progress.

The perceived precision of science ushered in the modern era, a socio-cultural movement largely characterized as a shift towards capitalism and secularism. From this pervasive modernist perspective, anything with the slightest hint of human subjectivity became suspect, neglected or, even worse, criticized as being a hindrance to this analytical understanding of progress. Human experience became utterly trivialized as human labor, as the sole industrial purpose of humanity, became merely a quantifiable means to a quantifiable end. This orientation towards modernist assumptions occurred roughly four centuries ago and has largely ingrained itself into society as an unquestioned truism. Yet, progress is something defined and determined by humanity, and is inseparable from our understanding of importance; that is, what is deemed important is the determining factor for our understanding of progress.

Alfred North Whitehead's examination of importance has monumental implications concerning this paradigm shift and the consequent predicament in which religion finds itself.[1] *Importance* is understood as an individual's selective emotive engagement with their present immediacy that consequently imbues the universe with meaning and coherence. An individual views the world through the lens of emotive importance and never completely objectively as the modernist perspective suggests. By further developing the ramifications of Whitehead's concept of importance, it is able to (1) demonstrate how and why religion lost its appeal in modernity; and (2) offer religion a different method to re-appropriate its place in modernity through realizing the way in which importance operates and shapes human existence.

As a way of giving some concrete examples to understand this different approach to religion, I open the chapter with two Christian thinkers who embody this Whiteheadian message—one from the Wesleyan tradition and the other from my own Roman Catholic tradition. Michael Lodahl argues for the employment of scientific knowledge in order to supplement and further

---

[1] "Whiteheadian" and "Process Thought" are used interchangeably throughout the paper and refer specifically to the thought of Alfred North Whitehead as opposed to the latter development of Process Theology.

his understandings of his Wesleyan tradition. Pope Francis, on the other hand, highlights aspects of his Roman Catholic tradition to display a commanding religious response to pressing ecological concerns. Both Christian thinkers embody a progressive approach by thinking outside of the self-imposed confinements of the scientific and religious disciplines, thereby opening a common ground of discourse revolving around their existential and communal worth.

As a brief disclaimer: I have chosen to address "religion" collectively, instead of just Christianity, because as the industrial complex continues to expand its influence throughout the world, all religions will eventually have to respond in one form or another. It is my hope to resonate with practitioners and believers of any particular faith community vexed by modernity's misappropriations of religion. Yet my examples throughout the paper are drawn exclusively from Christianity—the lived tradition I am most familiar with. In any case, each religious tradition will need to decide the best course of action, but I firmly maintain that Process Thought offers the most promising appropriation and response.

**Conveyors of Religious Worth**

To set the tone for our two contemporary Christian thinkers, a quote from Whitehead will effectively open the discussion to follow. In discussing the necessary conditions for generating a healthy worldview, Whitehead refers to the indispensable role of imagination. The imaginative twist, he suggests, enables humanity "to construct an intellectual vision of a new world, and it preserves the zest of life by the suggestion of satisfying purposes."[1] The meaning of life never suggests itself but instead humanity is to construct a worldview that serves an emotive purpose. The word *imagination* is less than favorable for more traditional accounts of religion, but it alludes to a crucial component of Whitehead's cosmology: creativity. Imagination or creativity need not be something mythical, fanciful, or utterly surreal. It is rather a critical engagement with any given *status quo* thereby giving new insights to past or contemporary issues that lead to new actions and experiences.

Some theologians and scholars of religion are taking this call to re-envision their religious perspectives very seriously. Within the Wesleyan tradition, Michael Lodahl argues for the utilization of different modes of knowledge to engage Protestantism. In his book, *God of Nature and of Grace,* he looks to the scientific disciplines to enhance his understanding of his faith. Instead of holding them in contention, he embraces the insights they have to offer one another:

> But the evidences of the natural sciences provide us radically different, new ideas about the world that have become a part of our experience of the world. This experience of the world in which we live provides an unavoidable and critical hermeneutical context for reading Genesis in a Wesleyan way.[2]

---

[1] Alfred North Whitehead, *Aims of Education and Other Essays* (New York: Free Press, 1967), 93.

[2] Michael Lodahl, *God of Nature and of Grace: Reading the World in a Wesleyan Way* (Nashville: Kingswood, 2004), 62. I would be amiss if I did not mention Thomas Jay Oord as another exemplary Wesleyan scholar being influenced by and implementing Whiteheadian insights to engage his faith. E.g., *Defining Love: A*

By putting these two different modes of knowledge into conversation with one another, he finds a fuller expression of not only scripture but human experience. The chasm between the two, Christianity and science, exhibit resonating factors when united in the sphere of human experience. The resultant effect is the creation of a re-envisioned world that shines new notions of importance upon the universe.

Concerning those who hold fast to *sola scriptura* and consequently denounce approaches that treat the "profane sciences" as a conservation partner within the Wesleyan tradition, Lodahl demonstrates that John Wesley himself integrated the science of his day and his understanding of faith. Wesley's sermon entitled "What is Man?" is a testament to this essential fusion.[1] This non-biblical addition to Wesley's thought shows his experiential attitude towards approaching scripture and tradition. Ingenuity such as this led to the formation of a prominent Protestant denomination. Interestingly, if John Wesley, the founder of Methodism, implemented religious investigations into scientific thought to supplement his faith, how, we may ask, did this kind of engagement become taboo in our present context?

Religions, therefore, have much to benefit from entering into the conversation with scientific thought as Lodahl suggests, but the benefit is one of mutuality. Whitehead goes even further in suggesting that science and religion have a reciprocal appeal to one another. In *Science in the Modern World,* he suggests that a unitive

---

*Philosophical, Scientific, and Theological Engagement* (Grand Rapids: Brazos, 2010).

[1] Lodahl, *God of Nature and of Grace,* 62.

harmony of contrasting experiences is manifest between science and religion:

> Science is concerned with the general conditions which are observed to regulate physical phenomena; whereas religion is wholly wrapped up in the contemplation of moral and aesthetic values. On the one side there is the law of gravitation, and on the other the contemplation of the beauty of holiness. *What one side sees, the other misses.*[1]

Religions and science, in this sense, have an *obligation* to engage one another in the discourse of modernity in a relational manner in order to expand human experience. Whereas Lodahl shows the potential benefits for religion seriously engaging scientific thought, our next Christian thinker embodies the great aesthetic awareness religion can add to scientific sensitivities.

Pope Francis is exemplary in this aesthetic consciousness engaging science.[2] His encyclical *Laudato Si'* is a prime example of the Roman Catholic tradition contributing its understandings to modernity. By bringing religiously grounded responses to the forefront of contemporary problems, Christianity becomes another voice alongside science and other disciplines. He looks to his tradition in hope of shedding new light unto the ecological crisis and finds St. Francis of Assisi as a righteous figure bursting with potential significance. The

---

[1] Alfred North Whitehead, *Science and the Modern World* (New York: Free Press, 1967), 185. Emphasis mine.

[2] The connection between the Pope's Encyclical and Process Thought has been discussed in more detail elsewhere. See John B. Cobb and Ignacio Castuera, eds. *For Our Common Home: Process-relational Responses to* Laudato Si' (Anoka: Process Century, 2015).

importance he finds is the applicability of St. Francis' message to ecological concerns.

Whereas science will have its own rhetoric, Pope Francis convincingly shows the communal worth of the Catholic saint in conversation with ecological concerns. Pope Francis claims the founder of the Franciscan order "shows us just how inseparable the bond is between the concern for nature, justice for the poor, commitment to society, and interior peace."[1] This statement is powerful in illustrating the interrelated issues involved in the ecological crisis. It is not merely a crisis of pollution and sustainability (the bare facts), but it is grounded in a capitalistic worldview leading to the impairment of nature, justice, and the poor (values). These keen discernments, arguably, would not have been acknowledged by other compartmentalized disciplines grounded in an industrial definition of progress. Demonstrating this type of sensible applicability to contemporary issues, it seems only natural that a resurgence of the interest will be brought back to Christianity as a supplementing or alternative voice to any given area of study.

These Christian thinkers and their potential success outside religious circles can be understood through an investigation into Whitehead's notion of importance. Importance never suggests itself but rather is generated and maintained within the sphere of human activity. By crossing the supposed bifurcated lines between religious knowledge and scientific knowledge,

---

[1] Pope Francis, "*Encyclical Letter* LAUDATO SI' *of the Holy Father Francis on Care for Our Common Home*," May 24, 2015. http://w2.vatican.va/content/francesco/en/encyclicals/documents/papa-francesco_20150524_enciclica-laudato-si.html, §10.

greater conformity to human experience is realized. These Christian thinkers conjoin contrasting worldviews to generate a more responsive intellectual vision of the universe. These theologians show a promising path for the future of religious discourse by engendering a space that closes the perceived chasm between science and religion.

**The Importance of Whitehead's Importance**

Process Thought has much to clarify in the preceding section. Whitehead's examination into importance implicitly evinces how these approaches can play a progressive and constructive role for religion in modernity. Importance plays a pivotal role by accounting for the way in which individuals and societies interact with their communities and the universe. Yet this topic seems to have never been fully developed as a referential tool for the religion-science conversation. It is a crucial investigation because importance (*un*)consciously operates in humanity's everyday evaluations and interactions within the universe. Whitehead creates a cogent cosmological vision that integrates the experiential aspects of existence with the physical data of the universe. Understanding the balance between these two components of reality is a crucial factor for Whitehead's paradigm. Reality, in its totality, is the intermingling of the two aspects and understood as distinctive yet unified features. Nowhere is this interaction more apparent than in the engendering of importance.

In this exploration, the first question to ask is how exactly does importance operate? Whitehead recognizes two determining factors in the generation of importance. In *Modes of Thought* he writes, "there are two aspects to importance; one based on the unity of the Universe, the other on the individuality of the details. The word interest suggests the latter aspect; the word importance leans towards the former."[1] The term *interest* refers to the intensity of an individual's feelings, which leads to a particular engagement with the universe. The integration of individual interest is implicitly invested into the unity of the universe and necessarily leads to the generation of importance and an ensuing worldview.

Importance, according to Whitehead, manifests itself as a practical perspective. Perspectives are importance in action and the interest of the individual linked to the unity of the universe. Put differently, importance is an abstracted view of the universe in the purview of an individual or society. Despite this worldview being lived out in the universe, Whitehead is quick to add that this manifestation of importance and the subsequent perspective are "*unexpressed* presuppositions, expressing the patterns of perspective as observed by the average human beings."[2] As long as an individual's perspective is reinforced, the individual's interest will arguably be unaltered and unquestioned. These perspectives become first nature to humanity—as opposed to second nature—and become the very lens through which humanity approaches and understands the universe.

A sensible perspective takes into account not merely the physical world but all the visceral intensities of experience. In *Process and Reality*, Whitehead shows the interrelatedness of morality, perspectives, and

---

[1] Alfred North Whitehead, *Modes of Thought* (New York: Free Press, 1968), 8.

[2] Whitehead, *Modes of Thought*, 11. Emphasis mine.

importance as it relates to an individual's conception of the universe. He states, "the *selectiveness* of individual experience is moral so far as it conforms to the balance of importance disclosed in the rational vision."[1] This rational vision consists of three tiers of reality: the Self, the Other, and the Whole. The divisions between these tiers are not clear-cut and they promote a holistic approach to the universe despite being loosely demarcated for conceptual reasons. Whitehead states, "it is the importance of the others which melts into the importance of the self."[2] Each tier is pivotal, yet an individual's moral selectiveness determines the importance that each tier receives. This selectiveness in experience is what Whitehead refers to as positive and negative prehensions, which operate consciously and unconsciously.

Whitehead's discernments in this first exploration highlight two crucial points for our conversation, namely, that importance is *created and selected* in producing a rational vision. The first point insists that importance is *not* a universal truth that humanity taps into or comes to know, but rather is developed in a relational character between the individual, the other, and the universe. Furthermore, it is moral insofar as it entails a selection or an abstraction of the whole as determined by the individual. The individual's understanding of importance is projected upon the universe thereby coloring the universe with her interests. Importance or interest, therefore, determines

---

[1] Alfred North Whitehead, *Process and Reality: An Essay in Cosmology* (New York: Free Press, 1978), 15. Emphasis mine.

[2] Whitehead, *Modes of Thought*, 117.

an individual's engagement and understanding of universe.

The second question, then, must address how importance and interest are determined for an individual. As addressed above, interest is never formed in a vacuum, nor is it ever a "given" within the universe. The interpenetration of these three tiers occurs continuously and seamlessly, yet, as Whitehead noted, the final decision is dependent upon an individual's own self-determinacy or subjective aims. Here, Whitehead's theory of language becomes indispensable in discussing the generation of importance/interest—more specifically, his understanding of a proposition. To be cognizant of something interesting is to be moved by a visceral feeling. As Whitehead states, "the primary function of a proposition is to be relevant as a lure for feelings."[1] The problem brought about by the analytical tradition is that it has limited the role of propositions to merely true-false assertions. The widespread influence of this restrictive understanding of propositions has resulted in making the primary function of propositions as lures for feelings to "fade into the background."[2]

By reinstating the primary function of a proposition, religion and forms of artistic expression can play a powerful role in a constructive postmodernity. Whitehead did not develop a theology, but he was acutely aware of the powerful expressions found within religious traditions and their inherent potential for transforming an individual's sensitivities towards greater harmony. This realization is culminated in his following insight:

---

[1] Whitehead, *Process and Reality*, 25.
[2] Whitehead, *Modes of Thought*, 37.

> [C]onsider a Christian meditating on the sayings in the Gospels. She is not judging 'true or false': She is eliciting their value as *elements in feeling*. In fact, she may ground her judgement of truth upon this realization of value.[1]

These lures for feelings are grounded in the intermingling of objective data from the historical and personal past of an individual while simultaneously being infused with the individual's subjective or emotional engagement with these data in their present immediacy.

The greatest function of religion, through the conduit of feelings, is to lure humanity beyond the physical facts of the universe in order to shape humanity's interest and find a satisfying emotive purpose. In other words, strong feelings provoke humanity to action, not facts alone. This insight is logically convincing when one understand that Whitehead propounds an experience based ontology as opposed to substance based ontology. If the world is constituted by different intensities of experience and events, feelings become powerful agents in reality. It is crucial to remember that these inspired purposes formed by propositions cannot be detached from humanity's experiences of the physical universe because without the physical universe to ground such purposes, the feelings would not arise in the first place.

Whitehead's insights into the nature of importance and interest, thus far, are simple yet brilliantly complex. The problem that arises is that humanity takes for granted the way in which we actively

---

[1] Whitehead, *Process and Reality,* 185. Emphasis mine. Pronouns changed to be gender inclusive throughout.

appropriate and participate in aspects of importance. Whitehead puts it succinctly when he states, "familiar things happen, and humankind does not bother about them. It requires a *very unusual mind to undertake the analysis of the obvious*"[1]—a very unusual mind indeed. This general investigation into the nature of importance reveals solid grounding to critique not only the overemphasis on science but also the problematic tendencies of religions, namely, being obsessed with the secondary, analytical understandings of propositions and its reluctance to constructively engage modernity.

**Importance and Religion in Postmodernity**

With some of the more theoretical and speculative notions behind us, a closer examination of importance in conversation with the scientific worldview and religion may ensue. The most intriguing insights from Whitehead's exploration shed light on two crucial details. First, *subjective* presuppositions ground the supposed objective or scientific worldview and, secondly, importance *undermines* truth. Each one of these insights warrants further discussion. Regarding the former, he writes:

> The most ardent upholders of objectivity in scientific thought insist upon its importance. In truth, "to uphold a doctrine" is itself such an insistence. Apart from a feeling of interest, you would merely notice the doctrine and not uphold it. *The zeal for truth presupposes interest.*[2]

---

[1] Whitehead, *Science and the Modern World,* 4. Emphasis mine.

[2] Whitehead, *Modes of Thought*, 8-9. Emphasis mine. This statement should not be misconstrued as positing a type of idealism.

Much to the dismay of those who uphold and praise the supposed value-free scientific understanding of universe, it is evident—within a Process worldview—that even the most unashamed objective inquiry is *riddled* with interest. Interest, as we may recall, is the individualistic rendering of importance. In fact, since science is a human enterprise, interest and value judgments are inescapable tangents of the scientific discipline. This being the case, humanity should dread the fact that our *selective* and *moral* notions of importance have developed atomic, biological, and chemical weapons whose sole purpose is the mass destruction of life. Returning to my opening paragraph, we must ask ourselves: Is this humanity's understanding of progress?[1]

Undoubtedly, the subjective interest of dogmatic objectivity could be a rallying call for other human enterprises that have been disenfranchised in wake of the modern scientific worldview, but this is a futile argument, which merely describes the current situation religion finds itself. Furthermore, the claim to the subjectivity of the sciences has the potential to relativize all points of view. Such an argument cannot be embraced by religion because it lacks something truly constructive. Whitehead's discussion concerning the relationship between truth and importance, in contrast, does offer the world's religions a constructive engagement with modernity.

---

Humanity does not determine truth, but humanity does determine which truths are important.

[1] This is not minimizing the great advancements within the scientific fields that have led to a better quality of life, but the destructive "advancements" in science are utterly frightening.

It is logically assumed that truth necessarily leads to importance, but Whitehead disagrees. Truth does not naturally equate to importance. He maintains that not only do we investigate our perspectives through our interest, but we always search for truth within the limitations of our interest.[1] Put differently, truth is contingent upon interest and directly leads to the self-inflicted problem facing religion in the contemporary world, which may be described as such: Any religious individual may argue that *truth* is clearly pronounced in the scriptures, the Dharma, or so forth, and constitutes an unalterable fact of reality, *but without interest* being generated by those pronouncements, humanity "would merely notice the doctrine and not uphold it," that is, humanity would notice scripture, the Dharma, and so forth, but not adhere to it.[2]

The lack of interest generated by religions is the crux of the problem. The importance-truth relation is able to vividly account how and why religion has lost its influential voice in modernity. The scientific worldview came to prominence, bringing its narrow yet practical worldview, when religion refused to dialogue with other developing modes of knowledge. The Western world is guilty of such and Christianity is a prime example. When Christianity went on the defensive in the face of the changing world initiated by science and its varying disciplines, it first entered into a self-inflicted stagnation. The Christian majority responded by clinging to and hoping to restore the past. Secondly, and more damaging, creeds became dogmatic expressions of faith instead of embodying the existential experiences that

---

[1] Whitehead, *Modes of Thought*, 15.

[2] Whitehead, *Modes of Thought*, 8-9.

originally inspired their formulation, a point Whitehead recognized: "The dogmas of religion are the attempts to formulate in precise terms the truths disclosed in the religious experience of humankind."[1] Yet, in the wake of modern analytical tradition, propositions became either true or false whereby situating Christian scripture in a foreign domain for which these religious lures for feelings were never intended. Religious expressions are not to be simple axiomatic proclamations of faith but a call to the immediacy of experience. Christianity isolated itself in order to protect itself from the changing times, but this approach backfired.

The isolation tendency is still arguably religion's *modus operandi*. The perceived fear of different modes of thought is kept at a distance, allowing religious communities to function as they please; however, this type of isolation equates to irrelevancy. The way to bring religion under the realm of importance once again is contingent upon religions' revision of their scriptures and traditions in light of the whole strata of human existence, namely, knowledge and experiences. Scripture, the Dharma, and other disciplines are resources for religions to draw from with an imaginative twist, as was demonstrated through Lodahl and Pope Francis. We cannot change the past, but we can breathe life into it by employing an array of different lenses grounded in contemporary understandings and lures for feelings.

Religions must take responsible for their irrelevancy within our present context. While scientific thought flourished and continues to flourish into all

---

[1] Alfred North Whitehead, *Religion in the Making* (New York: Meridian Books, 1960), 47.

facets of human life, the general response from religions was to live on the nostalgia of the past, digging their heels, and wait for the world to come to its religious senses. Whitehead's theory of importance and its relation to truth crystalizes religion's irrelevancy in modernity. Something more creative and captivating is needed. This Whiteheadian investigation into importance problematizes religion's role in postmodernity and additionally shows possible avenues to reconcile the fractured relation.

**ReGenerating Importance**

The new emphasis for religion in this Process exploration squarely establishes potential paths for its engagement with modernity. The Whiteheadian path is based upon establishing paradigms of interest-importance (Constructive Postmodern approach) and not being preoccupied with the traditional pronouncement of truth claims (Modern Analytical approach). Whereas the former constructively intensifies individuals' emotive sensitivities, the latter essentially falls upon deaf ears within religious discourse. The objective for religion is to have humanity not only notice a doctrine but uphold it through generating importance. This insight shifts the focus internally onto religions themselves, thereby leading to new characterizations and re-initiating their engagement in postmodernity.

Religion must find its place alongside science and other disciplines by demonstrating its acute worth. The ultimate goal of recaptivating interest in religion is not to usurp the position of science, but rather balance the pendulum of what constitutes human experience—a mean between the extremes of objectivity and subjectivity. The beautiful harmony surrounding

Whitehead's cosmology attempts to account for all experiences derived within and through the universe (human or otherwise) by stressing the inescapable relationality of all things. The way to generate importance is contingent upon understanding the contrastive (not oppositional) nature of different modes of knowledge and experiences that ultimately find a unitive element through their existential and communal importance.

A detached scientific perspective trivializes the rich fullness of the human experience by treating humanity as a mere spectator of the universe as if human experience and value are meaningless. As Whitehead explicitly states in *Principles of Nature Knowledge*, "life is complex in its expression, involving *more than percipience*, namely desire, emotion, will, and feeling. It exhibits variations of grade, higher and lower, such that the higher grade presupposes the lower for its very existence."[1] Religion's purpose is to elicit interest through the intensity of feelings. These feelings can frame an individual's truths, but they must be feelings before they are lived truths. The ultimate goal of religion is to shape our approach to the self, the others, and the universe, and not describe barren facts of the universe. It is a categorical mistake to view religion in terms of the latter. Yet, once interest is cultivated, new feelings and sensitivities will naturally come into focus.

Again, this is not to say that religion is to trump science. The scientific and religious perspectives are essentially monotone if and when they neglect other

---

[1] Alfred North Whitehead, *An Enquiry Concerning the Principles of Natural Knowledge* (Cambridge; University Press, 1925), 197. Emphasis mine.

experiential intensities that constitute *lived* life. Relatedness, either epistemic or ontic, is an ultimate ontological necessity. Whitehead had an aversion to any philosophy propounding rigid demarcations because knowledge and our engagement with the universe are never experienced as such. In *Religion in the Making,* he argues for a fuller engagement between experience and knowledge because they can never be completely compartmentalized from one another:

> Religion, therefore, while in the framing of dogmas it must admit modifications from the *complete circle of our knowledge*, still brings its own contribution of immediate experience.[1]

By working with the complete sphere of human knowledge and experience, unique insights that are otherwise neglected in narrow perspectives come into focus.

**Closing Remarks**

I have concluded that the failure of religion is its refusal to reimagine itself within newly emerging disciplines—scientific or otherwise. These other disciplines creatively and constantly re-envision the universe and subsequently inspire humanity's creative impulses and imagination. Religion, on the other hand, fails to stimulate humanity to the same degree as before because of its inability to look outside of its normative, analytical understandings in order to reappropriate itself in the context of postmodernity. Furthermore, Whitehead's examination into the relation between importance and truth reveals the unexpressed values of modernity, leading to a possible critique in favor of

---

[1] Whitehead, *Religion in the Making*, 67. Emphasis mine.

religions by demonstrating that strict objectivity or "value-free scientific inquiry" is nothing more than a popularized and persuasive myth. Modernity has taught humanity to be uncritical of "value-free" notions or, put differently, modernity has taught humanity to embrace "value-free" values. However, as I argued above, this critique is futile because it merely asserts humanity's predicament without giving it a way out. Religious sensitivities, however, can bring a more nuanced balance to humanity's values.

The Whiteheadian paradigm of the universe is one that is based on importance as opposed to the bare, cold hard facts. This discernment problematizes modernity's current overemphasis on the scientific worldview, but it also suggests an avenue to respond constructively. The task is challenging but decisive for religions—reinvigorate interest and importance though demonstrating their existential and communal worth by relating their particular perspectives to the wider discourse of knowledge and experience. This discourse is to be anchored upon the aesthetic, existential, and communal characteristics of religion, and not based upon the appeal to revelation alone. Religion must lure humanity to new values, not tell them new values. The importance of a perspective is generated in a dynamic relational character and determined by its potential applicability and inspirational qualities to other modes of thought outside of its own normative understanding.

A meaningful cosmological vision should attempt to incorporate as many facets of experience and fields of knowledge as possible. Selectiveness is unavoidable in this process, but wider breadth equates to greater interest and importance. The "sacred" cannot proclaim itself as something unrelational, as something "not of this world" in the face of the "profane." Segregation leads to irrelevancy and conversation leads to importance. As Whitehead warned, "apart from a *feeling of interest*, you would merely notice the doctrine and not uphold it."[1]

The study of Michael Lodahl and the proclamations of Pope Francis address this failure head-on whereby exposing new modes of dialogue with modernity by appropriating its knowledge and not shying away from it. They break-out of compartment-alized or normative modes of knowledge and demonstrate the fruits of utilizing knowledge in a more unitive sense. They engage scientific inquiry as a partner in addressing different topics and do not attack it as an epistemic enemy. These two Christian thinkers are some of the most recent examples, but it is certainly not limited to them. Three centuries ago, John Wesley used an array of *secular* knowledge to inform and color his sacred theological pronouncements. The opinions differed between the disciplines but all of them shared an existential concern. For John Wesley, the common ground was the heart: "Though we cannot think alike, may we not love alike? May we not be of one heart, though we are not of one opinion? Without all doubt, we may. Herein all the children of God may unite, notwithstanding these smaller differences."[2] Although this statement was directed towards his Roman Catholic brethren, it can also be applied to scientific disciplines. This resonates with Whitehead's assertion that "a clash

---

[1] Whitehead, *Modes of Thought*, 8-9. Emphasis mine.

[2] John Wesley, *John Wesley* (ed. Albert Cook Outler; New York: Oxford University Press, 1964), 93.

of doctrines is not a disaster—it is an opportunity."[1] The opinions differ between "the sacred and the profane," but the importance of creating a harmonious worldview remains the same.

In the end, we realize humanity cannot be mere bystanders to the physical laws of the universe, because humanity's intrinsic nature is to breathe purpose and meaning into it. But human beings can neither solely ground themselves on a religious tradition because "churches, rituals, bibles, codes of behaviour, are the trappings of religions, its passing forms.... But the end of religion is beyond all this."[2] Religion is to be more than "tradition" but continually lure humanity to new becomings in the face of the unknown future. As different modes of thought are developed and furthered in modernity, religion must constantly develop and further itself. This sentiment is captured by the Whitehead scholar Roland Faber, who contends that religion is "a continuity of becoming and not an identity of an essence."[3] Truth is everywhere, but the shifting notions of importance determine if this or that truth is acknowledged on a wide scale. A truth within itself cannot generate interest or importance, and it is exactly at this point that the undeniable role of humanity is clear according to Whitehead: Humanity appropriates truth through importance. The future of religion depends on its ability to intensify human feelings manifesting in interest and importance, thereby liberating the notion of truth.

---

[1] Whitehead, *Science in the Modern World,* 186.

[2] Whitehead, *Religion in the Making,* 17.

[3] Roland Faber, "Der transreligiöse Diskurs. Zu einer Theologie transformativer Prozesse," *Polylog* 9 (2002): 79.

# *Panentheism*

## *A Potential Bridge for Scientific and Religious Dialogue*

*Joyce Ann Konigsburg*

One of humanity's ultimate questions that intrigues science, philosophy, and theology is the origin of the cosmos. During interdisciplinary dialogical encounters, scientists theorize about a "big bang" that initiates a chain of reactions forming the universe while theologians focus on the Creator's relationship with creation. A variety of approaches exist to analyze the doctrine of God in relation to the cosmos. At one end of the spectrum, pantheism considers God and the cosmos to be equivalent; in fact, so co-dependent that God *is* the universe. At the other end, classical theism views the relationship as discrete; the infinite Creator is present yet distinct from and thereby unaffected by finite creation. In between these perspectives is panentheism, which translates to "all-in-God."[1] Panentheism claims God encompasses the world in a reciprocally influential relationship, however, God also transcends it.

Many Christian sermons and writings contain panentheistic concepts. For example, John Wesley says:

[1] The early 19th century German philosopher Karl Christian Friedrich Krause first used the term panentheism. For more information, refer to Joseph A. Bracken, "Panentheism in the Context of the Theology and Science Dialogue," *Open Theology* 1 (2014): 1–11.

> God is in all things, and… we are to see the Creator in the glass [sic] of every creature… who by his intimate presence holds them all in being, who pervades and actuates the whole created frame, and is in a true sense the soul of the universe.[1]

Wesley's actual meaning is subject to hermeneutical interpretations about what the word "in" signifies for the God-world relationship. Likewise, Pope Francis writes in his encyclical, *Laudato Sí*, "The universe unfolds in God, who fills it completely. Hence, there is a mystical meaning to be found in a leaf, in a mountain trail, in a dewdrop, in a poor person's face."[2] These statements about creation and the environment also might be interpreted as allusions to panentheism. If a pope references mystical, panentheistic ideas inspired from liberation theology, his namesake who is the patron saint of animals and nature (Francis of Assisi), and the Sufi mystic, Ali al-Khawas,[3] perhaps it is time to evaluate panentheism as a potential bridge for scientific and religious dialogue.

**Representative Approaches**

Though experiencing a resurgence in popularity, notions of panentheism occur in ancient Egyptian poetry, Hinduism's sacred writings, Greek philosophy, as well as Jewish, Christian, and Islamic Scriptures. In an

---

[1] John Wesley, *The Works of the Rev. John Wesley: Forty-Two Sermons on Various Subjects* (New York: J & J Harper, 1826), 5:243.

[2] Pope Francis, *Laudato Sí* [On Care for Our Common Home], Vatican Web Site, (May 24, 2015): http://w2.vatican.va/content/francesco/en/encyclicals/documents/papa-Francesco_20150524_enciclica-laudato-si.html (accessed February 1, 2016), para. 233.

[3] Pope Francis, *Laudato Sí*, n 159.

early Vedic Hindu myth, the gods sacrifice Purusha's cosmic body to bring the world into being and then sustain it.[1] Lord Krishna expresses panentheistic qualities in the *Bhagavad Gita,* when he declares, "with a single fragment of Myself, I pervade and support this entire universe."[2] Mystical texts in Judaism and Islam imply that God permeates the physical world. Hence, panentheism manifests within a variety of religious traditions. These sacred writings and philosophical ideas significantly influence classical theism, especially its understanding of the God-world relationship.

Christianity also espouses several panentheistic models and metaphors. Expressivist panentheism claims God is "a self-conscious subjectivity who creates the otherness of creation in order to bring it back into divine life"[3] through the Holy Spirit. Within Greek Orthodox traditions, soteriological panentheism is a gift; "the communitarian life of the three divine persons of the Christian doctrine of the Trinity is freely offered to all creatures at the end of the world."[4] Nevertheless, classical soteriology constrains such coexistence to only "those aspects of created reality that have become godlike, while they still remain a created reality"[5] until the consummation of creation at the eschaton. These two

---

[1] *Rig Veda,* http://www.sacred-texts.com/hin/rigveda/rv10090.htm (accessed 119/2016).

[2] *Bhagavad Gita,* 10.42, http://www.bhagavad-gita.us/bhagavad-gita-10-42/ (accessed 1/16/2016).

[3] Bracken, "Panentheism," 4.

[4] Bracken, "Panentheism," 1.

[5] Niels Henrik Gregersen, "Three Varieties of Panentheism," *In Whom We Live and Move and Have Our Being: Panentheistic Reflections on God's Presence in a Scientific World* (ed. Philip Clayton and Arthur Peacocke; Grand Rapids: Eerdmans, 2004), 21.

models of panentheism encourage unity and preserve difference in Creator–creation relations yet they avoid arbitrary notions of the world or any loss of its freedom.

Originating from process theology, dipolar panentheism asserts that God possesses seemingly contradictory attributes as well as both a primordial and a consequent nature. The dipolar understanding of God, for example, instigates an immutability-mutability tension. Because God is eternal potentiality as well as temporal actuality, God is the unifying principle for a world capable of change, diversity, and development.[1] According to Charles Hartshorne, the God-world relationship requires that "a supreme person [God] must be inclusive of all reality… since relations contain their terms, persons must contain other persons and things."[2] Thus, God is in the world and vice versa. Philip Clayton similarly favors internal growth and interdependence between God and the world that provides order and predictability to the cosmos.

Metaphors frequently describe panentheistic God–world relationships. Some examples include God as composer, artist, or ocean, with the world as music, art, or a sponge, respectively; but these models convey physically separate entities rather than the notion of being "in" each other. Depicting the world as a perpetually unborn baby in God's womb is a powerful analogy that combines separate entities, but it implies

---

1 Charles Hartshorne and William L. Reese, eds. *Philosophers Speak of God* (Chicago: University of Chicago Press, 1953), 273.

2 Charles Hartshorne, *The Divine Relativity: A Social Conception of God* (New Haven, CT: Yale University Press, 1948), 144; See also Owen C. Thomas, "Problems in Panentheism," *The Oxford Handbook of Religion and Science* (ed. Philip Clayton; Oxford: Oxford University Press, 2006), 655.

female divinity. Sallie McFague and Grace Jentzen revive the "world as God's body"[1] metaphor, which originates as far back as Stoicism, Platonism, and early Christian theologians like Tertullian and Irenaeus. This panentheistic metaphor serves as an alternative hermeneutic starting point to the prevailing traditional metaphor of God as King. The King metaphor encourages hegemonic, hierarchical, patriarchal thinking, while the world as God's body provides inclusive, equitable, non-dualistic language more compatible with feminist and ecological theologies.[2] However, a primary concern about the world as God's body is reductionism, which closely resembles pantheism. McFague argues that human bodies are expressions of the personal agents who reflect upon them; similarly, God's personal, creative agency prevents the world from being all that God is.[3] Furthermore, imagining the world as God's body affirms goodness and value in the cosmos. It also suggests dependence and vulnerability for the world without limiting God.

**Challenges to Orthodox Theology**

Classical theism characterizes God as absolute and supernatural, who freely creates then sustains the universe. Among the varying list of divine attributes,[4]

---

[1] Sallie McFague, *Models of God: Theology for an Ecological, Nuclear Age* (Philadelphia: Fortress Press, 1987), 61. For more information, refer to Grace Jantzen, *God's World, God's Body* (Philadelphia: Westminster Press, 1984), especially chapter 3.

[2] McFague, *Models of God*, 63–69.

[3] McFague, *Models of God*, 73.

[4] For general information on Western theism, refer to William Wainwright, "Concepts of God," *The Standford Encyclopedia of*

immanence (presence to creation) and transcendence (completely other) cause the most conflict with ideas found in panentheism. Michael Brierley identifies three general principles differentiating panentheism from classical theism. Succinctly, classical theists affirm 1) God is more than the cosmos, 2) God is not separate from the cosmos, and 3) God is affected by the cosmos.[1] Brierley's description is similar to Gregersen's depiction of generic panentheism in which "the cosmos is in some sense 'in' God and ... the relations between God and cosmos are 'in some sense bilateral.'"[2] These descriptions of panentheism affirm divine transcendence but challenge classical theism's concept of divine immanence. Despite divergent perspectives, both classical theism and panentheism must address the problem of evil, while notions of divine omnipotence and causality are contentious for science-religion dialogue.

**Immanence and Transcendence**

Within classical theism, divine immanence refers to God's nearness or presence in the material world, which often manifests in humanity's history. Panentheism describes divine immanence as a world-in-God and God-in-world bidirectional, interdependent association. Western Christian theology concurs with some of the God-in-world aspects of panentheistic immanence. Thomas Aquinas, for example, says that "God exists in everything... as an agent is present to that

---

*Philosophy* (Spring 2013): http://plato.stanford.edu/archives/spr2013/entries/concepts-god/ (accessed 2/1/2016).

1 Michael W. Brierley, "The Potential of Panentheism for Dialogue between Science and Religion," *Oxford Handbook of Religion and Science*, 641.

2 Gregersen, "Three Varieties of Panentheism," 22.

in which its action takes place.... So God must exist intimately in everything."[1] Classical theist Karl Barth acknowledges:

> God has the freedom to be present with that which is not God, to communicate Himself and unite Himself with the other and the other with Himself, in a way which utterly surpasses all that can be effected in regard to reciprocal presence, communion, and fellowship between other beings.[2]

If classical theism and panentheism agree that God either is or possesses the freedom and capability to be in the world, then they differ on the idea of all-in-God.

Traditionally, theological debates regarding panentheism focus on what the pivotal word *in* signifies for the God-world relationship. Given the diverse theological, philosophical, as well as grammatical interpretations of the word *in*, some panentheistic explanations favor pantheism while others coincide with attributes of Classical theology. Moderate panentheistic approaches resonate positively during scientific and religious dialogue. Nevertheless, the idea of all-in-God poses challenges to God's ontology, identity, and relationship to creation.

For classical theists, divine transcendence provides an essential ontological distinction between God and the world. Rather than profess that God is part of or dependent on the world, theological reflection on the Scriptures discerns a significantly radical concept of differentiating the Creator from creation. In what Robert

---

[1] Gregersen, "Three Varieties of Panentheism," 23; See also Thomas Aquinas, *Summa Theologiae*, I Q8.1.

[2] Karl Barth, *Church Dogmatics* (ed. G. W. Bromiley and T. F. Torrance; Edinburgh: T & T Clark, 1957), II/1:313.

Sokolowski calls the Christian distinction, "God is understood as 'being' God entirely apart from any relation of otherness to the world or to the whole. God could and would be God even if there were no world."[1] Asserting that the divine is wholly other ontologically differentiates God from imperfection, evil, and suffering in the world along with other limitations associated with the physical universe; yet God's immanence remains crucial in order to avoid dualisms equating good with spiritual and evil with material things.

Several forms of panentheism likewise distinguish between the divine and the world in order to retain their ontological uniqueness. Panentheism asserts that "while the universe is part of God, God and the universe do not form an undifferentiated whole."[2] God is total, inclusive, self-existing wholeness that is more or greater than the universe, though the created universe is whole unto itself. In dipolar panentheism, God and the world are dialectical opposites, so both entities are equally transcendent; thus "God transcends the World in that the World originates from within the divine Ground of Being."[3] Panentheism also considers many of classical theism's divine all-encompassing attributes of omnipresence, omniscience, and omnipotence as pertaining to the Creator, but not to the universe or individual

---

[1] Robert Sokolowski, *The God of Faith and Reason: Foundations of Christian Theology* (Washington, DC: Catholic University of America Press, 1995), 32–33.

[2] David H. Nikkel, "Panentheism," *Encyclopedia of Science and Religion* (ed. J. Wentzel Vrede van Huyssteen; New York: MacMillan Reference USA, 2003), 641.

[3] Bracken, "Panentheism," 5.

creatures.[1] If distinctness means having or lacking essential properties, then God possesses divine attributes the world lacks, which establishes at least a qualified ontological God-world distinction.

Ontological distinction has an additional benefit of establishing and preserving unique, separate identities for God and the world. Differentiation between identities creates otherness, which is a crucial component in relationships. From an epistemological viewpoint, God's distinction makes the divine unknowable from human experience unless human knowledge extends "to things presupposed to our experiencing anything at all"[2] by using structures of experience or consciousness. As a result, panentheism argues that divine transcendence in classical theism has a moral cost because the God-world relationship serves as an example for internal relationships within the cosmos.[3] Although classical doctrines of God require the Creator be distinct from creation, the cosmos actually is dependent on its Creator. God creates, sustains, and guides the universe toward its ultimate purpose.

**Problem of Evil**

Proper God-world distinction avoids good and evil sharing an equal ontological status with the divine. However, with its assertion of all-in-God, panentheism nullifies that distinction. Consequently, panentheism as well as classical theism must address the problem of evil.

---

[1] Nikkel, "Panentheism," 641; Note that process forms of panentheism find the notion of divine omnipotence to be problematic.

[2] David Burrell, *Faith and Freedom: An Interfaith Perspective* (Malden: Blackwell Publishing, 2004), 205.

[3] Brierley, "The Potential of Panentheism," 645.

Some panentheistic theologians agree with Augustine and Aquinas that evil, as the absence of good, has no existence *per se*.[1] To the contrary, David Ray Griffin believes evil exists within God, but "only in God's experience, not in God's intentions [or essence],"[2] so no moral evil is in God. Arthur Peacocke concurs that natural and human evil are internal to God, yet through divine free choice "God can thereby transform it into what is whole and healthy."[3] In equating the world to God's body, McFague theorizes that all evil, pain, and suffering experienced by creation is likewise experienced by God.[4] Unfortunately, "the God who suffers... cannot wipe out evil"[5] without the assistance of humans, who actively choose to sin or to eradicate wickedness. Process panentheism absolves God from responsibility for evil by denying divine omnipotence and emphasizing creation's free choices.

**Divine Action and Causation**

Many of classic theism's notions about divine action originate from Aristotle's substance ontology and causality, which require that a First Cause or Prime Mover of creation exist who is not composed. Thomas Aquinas argues that every created thing is a composition

---

1 Augustine of Hippo, *De Civitate Dei*, XII.6, 22; Aquinas, *Summa Theologiae*, I Q48.1, Q12.6–7.

2 David Ray Griffin, "Panentheism: A Postmodern Revelation," *In Whom We Live*, 46.

3 Arthur Peacocke, "Articulating God's Presence in and to the World Unveiled by the Sciences," *In Whom We Live*, 151–52.

4 McFague, *Models of God*, 75.

5 McFague, *Models of God*, 75.

of what it is (essence) and an act of existing (existence).[1] Since God is *ipsum esse subsistens* (subsistent being itself), Aquinas asserts the First Cause is God, who is the only absolutely simple, perfect, indistinguishable essence and existence.[2] The Creator's continual active causality creates and maintains the essential being of all existent things; therefore, because "being is innermost in each thing... it must be that God is in all things and innermostly"[3] present. As First Cause, God is transcendent and totally other to creation, yet through divine causality, God is immanently active in creation by loving and sustaining it.

Aquinas's implementation of First Cause leads to a seemingly deterministic worldview that appears to negate or deter human freedom when interpreted as an omnipotent God creating what Gottfried Leibniz calls "the best of all possible worlds."[4] Classical theists insist "the creator who is the cause of being is also an intentional cause,"[5] in support of a free creation. Hence, participation in divine being expresses creaturely freedom and creativity. In fact, "all that exists participates in its own way in divine being through the very gift of creaturely existence,"[6] which maintains integrity and decision-making freedom. As participants in creation, human beings exercise free choice along with

---

[1] For more information, refer to Thomas Aquinas, *On Being and Essence*, ch. 4.

[2] Aquinas, *Summa Theologiae*, I Q3.4.

[3] Aquinas, *Summa Theologiae*, I Q8.1.

[4] Gregersen, "Three Varieties of Panentheism," 29.

[5] Burrell, *Faith and Freedom*, 153.

[6] Elizabeth A. Johnson, "Does God Play Dice? Divine Providence and Chance," *Theological Studies* 56 (1996), 11–12; see also Thomas Aquinas, *Summa Contra Gentiles*, 3.69.15.

its associated responsibility; consequently, God does not compete with creatures. Instead, as an intentional agent, the Creator inspires personal relationships with creation. Creatures in turn recognize the source of their being and express proper gratitude to the Creator.[1] In short, freedom in creation, which includes humanity, manifests in its free response to God's gratuitous creativity.

Panentheism advocates indeterminism, spontaneity, and free will within the universe. Both God and creation exercise free choices but without antecedent causes influencing events or actions producing an unpredictable future.[2] In process panentheism, God presents possibilities to world events, then each event determines how to actualize those possibilities. This freedom, along with the absence of divine determination, also prevents God from being the direct cause of evil. Yet God loves, respects, and upholds human freedom "even at the cost of bringing evil into the world as an indirect consequence."[3] Complete freedom is necessary for humans to be fully in God on a personal level. In such relationships, God is "self-conscious subjectivity who creates the otherness of creation in order to bring it back into divine life."[4] Thus, inter-subjectivity goes beyond subjectivity in describing the actual God-world relationship.

Of mutual concern for scientists and theologians is divine action and its implications of intervention into the natural order. The scientific community is more agreeable to a Creator establishing universal laws

---

[1] Burrell, *Faith and Freedom*, 153.

[2] Nikkel, "Panentheism," 641.

[3] Bracken, "Panentheism," 4.

[4] Gregersen, "Three Varieties of Panentheism," 30.

(general providence) than God directly causing specific events in world history (special providence).[1] William Stoeger posits that "God is always acting through the deterministic and indeterministic interrelationships and regularities of physical reality which our models and laws imperfectly describe."[2] Classical theism proponents claim that even for divine acts perceived as extra-ordinary experiences (miracles), God "wills to activate and to sustain in act all those secondary causes whose activity contributes to the unfolding of the natural order which he [sic] intends to produce."[3] God's actions do not necessarily challenge physical laws or correct creaturely free will. Nevertheless, scientists perceive classical theism's notions of divine causality as intervention that disrupts nature's laws and contradicts logical theories based on experimentation. From the panentheistic world-in-God scenario, scientists consider divine action to be within the world's regular dimensions and thus not a violation of cosmic processes. Owen Thomas argues that in panentheism, God also is intervening from outside the world's closed causal nexus, although Thomas admits the closed model no longer reflects newer perspectives from quantum physics.[4] The point is

---

[1] Nikkel, "Panentheism," 642.

[2] William R. Stoeger, "Contemporary Physics and the Ontological Status of the Laws of Nature," *Quantum Cosmology and the Laws of Nature: Scientific Perspectives on Divine Action* (eds. Robert J. Russell, Nancey C. Murphy, and Chris J. Isham; Vatican City: Vatican Observatory Publications, 1993), 234.

[3] International Theological Commission, *Communion and Stewardship: Human Persons Created in the Image of God* (2004), http://www.vatican.va/roman_curia/congregations/cfaith/cti_documents/rc_con_cfaith_doc_20040723_communion-stewardship_en.html (accessed 3/13/2015), 66.

[4] Thomas, "Problems in Panentheism," 660.

that any intervention, whether internal (panentheism) or external (classical theism) conflicts with a closed causal world unless it is not as closed as scientists theorize.

**Advantages for Science-Religion Dialogue**

Challenges surrounding divine action and causation stimulate dialogue between scientists and theologians. Panentheism offers world-in-God models that facilitate traversing the physical and spiritual aspects of reality. Consequently, advocates of panentheism think it resonates with scientific endeavors much better than classical theism. In fact, many concepts about panentheism evolve within the context of science and religion discourse. Classical theism's emphasis on God as transcendent yet immanent implies an external relationship with the physical world; however, panentheism reflects the shift in science from a mechanistic to organismic worldview. Moreover, the panentheistic God-world relationship associates temporal characteristics with the divine that correlate with Darwin's ideas of evolution and Einstein's theory of relativity.

Though the relationship between science and religion has not been always beneficial or benevolent, many emboldened scientific discoverers, such as Galileo and Newton, pursued their work influenced by their deeply held religious convictions. Scientists often reconcile potential conflicts between such diverse passions by acknowledging, "God had written two books, the Book of Scripture and the Book of Nature. Both needed to be read, and when this was done aright there could be no contradiction between them, since the two had the same Author."[1] Contemporary scientists and theologians likewise recognize an increasing need for rapprochement since continued estrangement "leaves nature Godless and religion worldless [sic]."[2] Recent ecumenical efforts and interreligious dialogue encourage renewed collaboration between the two distinct disciplines. They believe that "science is the best available description of the physical world, and metaphysics should be practiced in continuity with physics (science)."[3] Moreover, both groups benefit from dialogue. Scientific method and physical evidence prevents religion from spiraling into superstition while religion addresses ethical issues that assist science in avoiding idolatrous pretentions of authority and false absolutism.[4] The introduction of scientific theories supports, constrains, or operates heuristically when developing theological concepts, while theology functions heuristically or recommends selection criteria when formulating scientific theories.

Existing differences between science and religion nevertheless affect their perspectives and methodologies. Both disciplines leverage valuable research and insight of previous experts; new scientific discoveries often result in discarding previous ones but religions must integrate new concepts or correct distortions within a

---

[1] John Polkinghorne, *Quantum Physics and Theology: An Unexpected Kinship* (New Haven: Yale University Press, 2007), 108.

[2] Robert Russell, "Bridging Science and Religion: Why It Must Be Done," *Theology and Science* 1.1 (2003): 1.

[3] Joseph Kaipayil, *Relationalism: A Theory of Being* (Bangalore: Jeevalaya Institute of Philosophy Publications, 2009), 40.

[4] Robert Russell, *John Paul II on Science and Religion: Reflections on the New View from Rome* (Vatican City State: Vatican Observatory, 1990), 13.

tradition's founding tenets. Scientists initiate new ideas about reality through observation and experimentation, while many religions believe creation discloses aspects its Creator or that the Creator reveals a hint of divine mystery through direct religious encounters, recorded within sacred writings. Repeatable scientific methods with verifiable experimental data foster greater general agreement whereas interreligious consensus is more difficult due to diverse cultural, metaphysical, and social perspectives, in addition to lingering adversarial historical relationships among faith traditions.[1] Confidence in the truths of science versus religion also varies; faith in scientific theories fulfills one's intellectual curiosity while religious faith entails personal and social ramifications beyond mere understanding.

Despite the many areas of dissimilarity between science and religion, both disciplines indeed search for truth about reality. As a result, John Polkinghorne envisions "a cousinly relationship between the ways in which theology and science each pursue truth within the proper domains of their interpreted experience."[2] He describes five points of this cousinly relationship by utilizing parallels between quantum mechanics and Christology. The first point comprises experiences of radical revision. Quantum mechanics struggled with wave-particle duality and early Christianity contended with Christ's divine-human nature. Therefore, the second point entails a period of unresolved confusion then a new synthesis and understanding, which is the

---

[1] One notable exception to this scientific unity is quantum theory with its multiple interpretations lacking clear consensus.

[2] Polkinghorne, *Quantum Physics and Theology*, 15. For more information on Polkinghorne's five points, refer to pages 15–22.

third point. Even with new understanding, the fourth point involves persistent unresolved problems until reaching the fifth point, an awareness of deeper implications that yield new insights for both disciplines.

Even though terminology and language problems create epistemological barriers to communication, panentheism utilizes scientific concepts such as supervenience, emergence, and quantum entanglement to foster understanding between scientists and theologians. Panentheistic scholars introduce theologies of divine action that cooperate with the laws of nature. Clayton and Peacocke believe no conflict exists between natural laws and divine agency; natural processes equate to God's creative acts.[1] Furthermore, the panentheistic world-in-God and God-in-world interdependence utilizes the scientific idea of strong emergence, particularly ontological emergence, which posits that reality entails one substance.[2] From the potentiality of objects, different properties emerge and develop into varying, hierarchically complex levels of supervenient systems with higher levels exerting causal influence onto lower ones "to suggest that God as a 'whole' influences or lures the 'parts' of the cosmos."[3] These internally related systems corroborate that "the world does constitute God's relational aspect but not the totality of God."[4] Thus, insights about the Creator-creation relationship suggest new models for trust and understanding in scientific and religious dialogue.

---

[1] Peacocke, "Articulating God's Presence," 144.

[2] Peacocke, "Articulating God's Presence," 137–42.

[3] Brierley, "The Potential of Panentheism," 643.

[4] Philip Clayton, "Kenotic Trinitarian Panentheism," *Dialog* 44.3 (2005): 250–54.

As an emerging theory of quantum physics, quantum entanglement offers promise for advancing science-religion dialogue. Modern quantum experiments demonstrate a physical relationality in nature previously not seen due to restrictions and isolation theories in Newtonian physics. Polkinghorne perceives entanglement as a "subtle form of interrelationality"[1] that he applies to the God–world relationship along with divine action and causality. Research by Thomas Tracy and Nancey Murphy substantiates "special providence by suggesting that God determines the probabilistic quantum movements of subatomic particles,"[2] which ultimately result in specific events without violating physical laws or scientific theories. Scientific theories about the natural world contribute to understanding divine action in the cosmos, but theologians and scientists should avoid rationalization about the God-world relationship or divine influence.

Because quantum physics reveals an indeterministic reality, scientists turn to philosophy for direction and insight, which provides another dialogic bridge with religion. Philosophical deliberations by scientists overlap traditionally religious questions about the origin and purpose of the universe, the nature of reality, and the existence of God. Whether or not scientific answers to these questions coincide with theological responses, the topics establish a common ground and shared interest upon which to engage in dialogue. Interestingly, scientific and religious dialogue

---

[1] John Polkinghorne, "The Demise of Democritus," *The Trinity and the Entangled World: Relationality in Physical Science and Theology* (ed. John Polkinghorne; Grand Rapids: Eerdmans, 2010), 7.

[2] Nikkel, "Panentheism," 643.

compels religion to reassess and reaffirm long-held beliefs. Interdisciplinary dialogue reiterates ontological, cosmological, and teleological arguments about God's existence along with God's active relationship with the physical world, either in a classical or panentheistic sense. Moreover, scientific observation, empirical methods, and logical analysis challenge theologians either to "emphasize the transcendence of God with renewed vigor and clarity or to abandon the doctrine altogether."[1] By participating in religious dialogue, scientists contribute new cosmological discoveries that augment traditional theological notions of creation and the eschaton. The result of science-religion dialogue is a resurgence in panentheism and natural theology that argues the existence of God based on reason and experience intermingled with orthodox religious beliefs.

Nevertheless, Neil Ormerod cautions scientific and religious scholars against conflating aspects of physics and metaphysics, especially in dialogue. He believes that as "fascinating as quantum mechanics is, the claims that insights into its account of physical phenomena give rise to a privileged metaphysical stance betrays an implicit metaphysical reductionism."[2] Likewise, empirical data that is fundamental to science perpetuates "what [Bernard] Lonergan calls the myth that reality is somehow 'already-out-there-now' waiting to be seen."[3] Such a myth limits physics and other

---

[1] Mark William Worthing, *God, Creation, and Contemporary Physics* (Minneapolis: Fortress Press, 1996), 202.

[2] Neil Ormerod, "Bernard Lonergan and the Recovery of a Metaphysical Frame," *Theological Studies* 74 (2013): 962.

[3] Ormerod, "Bernard Lonergan," 962. For more information, refer to Bernard J. F. Lonergan, *Insight: A Study of Human*

sciences to recognizing only visible aspects of the physical world. Metaphysics pertains to being as being, thus incorporating science into more all-encompassing intellectual endeavors. These expanded areas provide additional space for dialogue.

Another relevant topic for science–religion dialogue involves ethical issues resulting from advances in biology, physics, and chemistry. Human beings have a history of dominating the world by selfishly depleting its natural resources without considering how it affects other creatures. A panentheistic emphasis on all-in-God as well as metaphors of the world as God's body promote concern about the Earth and its environment. Recognizing the world as a unitary, interdependent whole also reinterprets the love of neighbor relationship to include respect for the Earth and its creatures. Christian social teaching likewise encourages stewardship and care for God's creation along with social justice, which are vital conditions for ecological integrity and harmony. Environmental ethics fosters genuine, altruistic concern for creation itself. However, ethics *per se* poses a problem for process panentheism. The notion of emerging complexity is too vague, too anthropocentric, and too dismissive, and it ignores or discounts 90 percent species extinction rates that evolution causes.[1] In addition, ideas of God and the world in the process of becoming, engender ethical relativism.

---

*Understanding* (ed. Frederick E. Crowe and Robert M. Doran; Toronto: University of Toronto, 1992), 276–77.

[1] Ruth Page, "Panentheism and Pansyntheism: God in Relation," *In Whom We Live*, 222–27; Thomas, "Problems in Panentheism," 662.

**Conclusion**

Panentheism offers several advantages to encourage productive scientific and religious dialogue. Imaginative metaphors of an intimate God–world relationship stimulate discussion between and among scientists and religious scholars. Expressing theological concepts utilizing scientific ideas and terminology such as supervenience, emergence, and quantum entanglement facilitates communication and increases understanding between the two disciplines. Brierley believes panentheism "represents theology that is engaging and creative, and a metaphysic with rich and sensitive moral resources"[1] able to address novel issues that science and technology introduce into society. Moreover, the increased popularity of panentheism provides nuanced alternatives to classical doctrines of God, which might mitigate historically religious opposition to science even as it dispels lingering apologetics and protectiveness.

Yet panentheism suffers from ambiguity regarding what it actually means for the world to be "in" God. In fact, Gregersen cautions: "panentheism should not be seen as a solution to the problem of thinking about God in a contemporary context"[2] because the concept lacks a clear definition that causes confusion and misunderstanding during interreligious and inter-ideological dialogue. Newer versions of panentheism encompass characteristics of divine immanence and omnipresence found in classical theism. Other models redefine or reinterpret classical doctrines to agree with

---

[1] Brierley, "The Potential of Panentheism," 648.

[2] Gregersen, "Three Varieties of Panentheism," 34.

panentheistic precepts. However, panentheistic views concerning interdependence and bi-directional influence between God and the world remain contrary to classical theism's ideas that the Creator is neither reliant on nor affected by anything. Dipolar process panentheism, for example, seems to contradict itself when describing divine attributes from two disparate perspectives. It also supports divine changeability, which creates ethical challenges of relativism during dialogue.

Terminology, metaphors, and theological subtleties associated with panentheism are more amenable to scientists in inter-ideological dialogue and facilitate respect, understanding, and cooperation. Although scientists value objective observation and experimentation of the physical world, they are also human beings who possess a variety of religious viewpoints. In addition, both theologians and scientists experience personal, subjective encounters with the divine. Often these encounters occur as scientists examine the world that theologians believe reveals its Creator. While proponents assert that panentheism functions as an effective bridge during science–religion dialogue because it is more compatible with science than classical theism, challenges may arise if panentheism causes cognitive dissidence with either the religious beliefs or the scientifically held theories of participants.

# *Divine Forgiveness and Mercy in Evolutionary Perspective*

*Isaac Wiegman*

In Abrahamic scriptures, God is often portrayed as both wrathful and just—a God who punishes the unrighteous and impure out of wrath and to satisfy the demands of justice. According to many in these traditions, the bad news is that all are unrighteous and all are impure, thus all deserve God's wrath and punishment. In many strands of the Christian faith in particular, this is the context in which the good news arrives: that someone else, a perfect substitute, has absorbed God's wrath or the punishment that humankind rightfully deserves or the death that is the proper repayment for human sin.[1]

Importantly, this cluster of views requires that God's wrath is *moral* wrath: wrath that is ignited by sin and that aims to consume impurities and satisfy the demands of justice. What kind of justice? On a natural reading of many texts, God's wrath appears to be fueled by *retributive* justice, as opposed to *restorative* justice or *distributive* justice. God's wrath is primarily aimed at giving sinners what they *deserve* irrespective of the overall *consequences* of inflicting that wrath (restoration, rehabilitation, deterrence, etc.). In other words, God's

---

[1] Among Abrahamic traditions, it is notable that Islam rejects substitution almost entirely. See e.g., John R. W. Stott, *The Cross of Christ* (Madison: Intervarsity Press, 1986), 45–46.

wrath is primarily a mechanism of debt collection rather than a mechanism for generating future returns. On this model, it is unavoidable that God's wrath is generated from within an economy of exchange within which each transgression is a debt that demands proportional repayment. This is what I will call *the system of payback*. Moreover, satisfying the requirement of payback via substitution must also be understood from *within* this system. The substitute pays the debt we sinners owe in accordance with the rules governing the system of payback.

To many, this understanding of humanity's relationship to God seems as natural and inevitable as the rising and setting of the sun. Nevertheless, closer scrutiny reveals deep mystery: by what moral alchemy does the suffering of the wronged coupled with the punishment of the wrongdoer transmute into a morally good occurrence? Why would God operate within such a system? Why would God be bound by the requirements of retributive justice? Why should God care about giving people what they deserve? At its root, retributive justice is concerned with proportional "payback" for transgressions or "balancing the scales." Despite the intuitive nature of these metaphors, it remains mysterious why anyone should care about payback or balance of this kind. It is hard to say why seeking payback is more reasonable than the alternative of letting bygones be bygones, leaving the past in the past, and instead promoting better outcomes in the future.[1]

---

[1] This is not to mention various puzzles about how desert claims could be satisfied by a substitute, since they ordinarily track individuals. See e.g. David Lewis, "Do We Believe in Penal Substitution?" *Philosophical Papers* 26.3 (1997): 203–09. See also Steven

While these are deeply interesting questions, my main aim is to tranquilize the mystery, not to give it teeth or defang it. In this essay, I explore the implications of prominent evolutionary explanations of payback, which exacerbate some of these questions. These explanations cast retribution as a product of the evolutionary pressures for self-protection that shaped our ancestors, rather than an immutable moral truth.[1] Moreover, the system of payback functions to constrain the negative effects of our evolved psychology. As a result, this scientific picture of human evolution may require a kind of Copernican shift in the way Christians understand the Good News of Jesus' atonement for human transgressions. The shift is away from a specific way of understanding Jesus' substitution for sinners in which the substitution is morally justified from within the system of payback. The shift is toward a view of substitution from outside the system of payback. On this view, the good news should not be understood as a moral transaction within it. Rather, our bondage to the system itself is the bad news, and the good news is that God has created an ingenious way out. Though the view from within the system may be as apparent as the rising of the sun, it may be equally inappropriate to take such appearances as reality.

---

L Porter, "Rethinking the Logic of Penal Substitution," *Philosophy of Religion: A Reader and Guide* (ed. William Lane Craig; New Brunswick: Rutgers University Press, 2002), 596–608; Mark C Murphy, "Not Penal Substitution but Vicarious Punishment," *Faith and Philosophy* 26.3 (2009): 253–73.

[1] "Immutability" is slightly misleading. I suspect that it would be more accurate to appeal to the notion of "stance-independence." See e.g. Russ Shafer-Landau, *Moral Realism: A Defence* (Oxford: Oxford University Press, 2003).

**Substitutionary Atonement**

Let us start with the bad news as many understand it.[1] Put simply, it is that "all have sinned," and "the wages of sin is death." Moreover, the traditional story goes that this debt is infinite in measure because our sins are against God who is infinitely good and these debts also separate us from relationship with God, which is the ultimate good for humankind. Finally, there is no way that sinners can pay for their sins or reconcile themselves to God through their own efforts. The debt incurred is infinite and our resources are finite, so we appear to be irretrievably mired in our debts.[2] The only way we can be saved is for someone to pay these debts on our behalf and for God to forgive our sins and restore relationship with us.

On this view, the good news is that Jesus has in some sense paid off our debt and made it possible for God to forgive our transgressions. In this matter, Jesus is our substitute, taking responsibility for the debts that we owe or absorbing the wrath or accepting the punishment that our sins deserve. Regardless of how substitution is understood (penal or otherwise), the underlying similarity is that Jesus takes responsibility for our sins and his suffering and death thereby absolve us from what we deserve or what we owe. On this cluster of

---

[1] Stott (*Cross of Christ*) gives a particularly clear and well-reasoned articulation of this view of the atonement and the scriptural case on which it is founded.

[2] Traditionally, the result of this is thought to be eternal condemnation in hell. For a discussion of divine retribution as it pertains to Hell, see Isaac Wiegman, "Divine Retribution in Evolutionary Perspective," *In Spirit and Truth: Philosophical Reflection on Liturgy and Worship*, (ed. Wm. Curtis Holtzen and Matthew Hill; Claremont: Claremont Press, 2016), 181–202.

views, Jesus' death is supposed to be a justifiable moral transaction.[1] Moreover, these views understand this transaction as necessary for either relational or impartial reasons. On the one hand, it is necessary to restore right relationship with God through forgiveness. On the other hand, it is necessary to change one's status, either through justification of some claim of justice or satisfaction of some moral claim concerning honor or desert.

**Relational Atonement and Forgiveness**

Consider the relational side first. Forgiveness is one way of restoring relationships after a transgression by relinquishing or revising one's attitudes toward a transgressor, attitudes such as anger, resentment, hurt feelings, or character evaluations. In the context of the atonement, Jesus death is supposed to be necessary for forgiveness in some sense. If so, it is natural to ask why God cannot just forgive us without payment?

The traditional response to this question is that forgiveness without payment would be inappropriate because of the seriousness of sin and because of the majesty of God. Such a response has a great deal of initial plausibility, especially when one considers recent philosophical discussions surrounding forgiveness. In these discussions, forgiveness is taken to be distinct from several other ways of revising one's attitudes (anger, resentment, etc.) toward a wrongdoer. As an example, consider that victims of abusive relationships will sometimes regulate their attitudes toward abusers by diminishing the abuser's responsibility for abusive

---

[1] I take it that the relevant cluster includes both penal substitutionary theories and satisfaction theories of the atonement.

actions ("she has been under a lot of stress lately") or accepting the abusive action ("I deserved that") or denying that the abuser's actions are abusive ("it didn't even leave a mark"). In none of these cases is the victim truly *forgiving* the abuser. As Pamela Hieronymi points out, one forgives under the following assumptions: (1) that the action was wrong, (2) that the agent was responsible, and (3) that the victim ought not to have been wronged. If this is right, then the seriousness of sin and the majesty of God are relevant to questions about forgiveness, because they reflect a basic commitment to the assumptions under which forgiveness operates. This puts a sharp point on the challenge of forgiveness more generally, which is to explain how one can change one's feelings toward a person while holding onto these assumptions.

Nevertheless, in ordinary cases of forgiveness, it seems perfectly consistent to forgive without proportional repayment. One way to think about this utilizes the metaphor of "wiping the slate clean."[1] When someone transgresses, their "slate" or record is marred by their transgression in that the transgression reflects on both their character and their relationship to the person they have wronged. When the wronged party wipes the slate clean, they make a choice to no longer identify the transgressor with the transgression. But if this view of forgiveness is correct, there is no obvious connection between forgiveness and payment. Though the action was wrong and the wrongdoer responsible for it (etc.), victims of transgression can still choose to break the connection between the transgression and their attitudes

[1] Lucy Allais, "Wiping the Slate Clean: The Heart of Forgiveness," *Philosophy and Public Affairs* 36.1 (2008): 33–68.

toward the wrongdoer without exacting any penalty. Apparently, one can make a decision and commit to the following view of the transgressor: "it was wrong that he lied and I deserved better, but he is not really a liar nor will I continue to resent him for it."

The traditional response could be built up in several different ways at this juncture to explain the necessity of repayment, but my interests here do not hinge on the specific ways this view could be developed.[1] The point I want to make is merely that appeals to the seriousness of sin and the majesty of God actually presuppose that payment is required for forgiveness. The weight of transgressions is irrelevant to forgiveness unless one assumes there is a scale that must be balanced before the slate can be wiped clean. As such, we can only answer the original question (of why God cannot wipe the slate clean without repayment) if we already have some reason to believe that such payment is required. But in discussions of this issue, such reasons are not forthcoming.[2] Rather, it is just *intuitively obvious* to some people that serious transgressions require some kind of repayment prior to forgiveness. Rather than rebutting various ways of defending this intuition, I suggest in section 3 that aspects of our evolved psychology allow us to *explain away* the necessity of repayment, by explaining this intuition.

---

[1] For a detailed discussion of this and related issues, see Lucy Allais, "Wiping the Slate Clean."

[2] See e.g., Stott, *The Cross of Christ*, chap. 4.

**Impartial Atonement and Penal Substitution**

In the domain of punishment and mercy, it is equally clear that we believe repayment is necessary in large part because it seems intuitively obvious.[1] We can see this by asking why punishment might be necessary to satisfy the demands of justice or to pay the debts that accrue to sin. This is an issue on which the Bible appears to remain almost completely silent. Instead, the Bible systematically *assumes* throughout that the righteousness or justice of God *means that* God will punish the wicked and reward the righteous. As Peter Enns notes, this is apparent in God's covenant with the Israelites (e.g., Deut 28) and in much of the wisdom literature of the Old Testament (e.g., Prov 3:33-35).[2]

There are a number of deeper reasons why justice might require the dispensing of rewards and punishments as "just deserts." I can think of only two that are relevant to substitutionary atonement.[3] One

---

[1] See for instance, Michael S. Moore, *Placing Blame: A Theory of the Criminal Law* (Oxford: Oxford University Press, 2010), chap. 4 for a defense of retribution that is ultimately based on intuition.

[2] Peter Enns, *The Sin of Certainty: Why God Desires Our Trust More than Our "Correct" Belief* (New York: HarperOne, 2016), 82.

[3] Four rationales for punishment are irrelevant here. First, God may threaten to punish the wicked in order to deter bad behavior. If this is the rationale, then punishment is a way of maintaining the credibility of the threat (see e.g., Warren Quinn, "The Right to Threaten and the Right to Punish," *Philosophy & Public Affairs* 14.4 (1985): 327–73.) But this reason for punishing actually has little to do with justice, since punishing an innocent can also maintain the credibility of a threat if everyone believes the person to be guilty. Second, God may punish to signal solidarity with the wronged or oppressed, but Jesus' punishment in place of the oppressor would do little to convince the oppressed that God sides with them *against* the oppressor. Substitutionary punishment is not well understood as an action that opposes an oppressor. Third, God might punish to

reason can only be explained by invoking a metaphor of balance. On this view, justice means trying to balance each person's consequences (i.e., rewards and punishments) with their actions. Perhaps one could claim that this kind of "karmic" balance is just a fundamental truth about the nature of justice which cannot be further explained.

But one might also attempt to give it a further explanation in terms of *goodness,* which brings us to the second reason. As Kant maintained, one might think that the "highest good" is "happiness in accordance with virtue," meaning that good outcomes would be apportioned to the virtuous and bad outcomes would be apportioned to the vicious. Why might this be the highest good? At this point, the lines of justification begin to "bottom out." For Kant, this is just a basic intuition about goodness. There appears to be little more to say than that it seems intuitive and fitting that virtuous people should get good things and that vicious people should get bad things.

---

communicate moral censure or disapproval of our actions (see e.g., Anthony Duff, *Punishment, Communication, and Community* {Oxford: Oxford University Press, 2001}.) and perhaps this is one reason why Jesus' punishment in our place might be effective. Nevertheless, this notion of substitution fits better with a moral improvement model of substitution than with the substitutionary views that are my focus here. The effect of punishment on this understanding would be an increased awareness of God's attitude toward our sin ("this is how much I disapprove of your sin"), rather than the kind of *justification* of the sinner that, for instance, penal substitutionary views attempt to capture. Fourth, God might punish to restore or enforce fairness, but it seems flatly contradictory to suppose that substitutionary punishment could be cast as restoring fairness. Surely punishing a substitute instead of a transgressor is a paradigmatic case of unfair action.

However, if we ask why this seems fitting, we might simply say that it is good to love the good and to hate the bad and that hating the bad (for instance) entails apportioning bad outcomes to people who are disposed to do bad things, namely the vicious.[1] But we can ask a still further question of why hating the bad requires giving them bad things. Why isn't it enough to simply prevent vicious people from doing bad things in the future and let the past remain in the past? It is easy to see that this line of questioning has gotten us no further in justifying the requirement that bad deeds or vicious people be repaid in suffering or hard treatment.

For my purposes here, what is important is that the main lines of justification for the necessity of payback appear to bottom out in intuition. In the following section, I make the case that our most basic intuitions on this matter probably derive from some of our most basic emotional responses which have been shaped by our evolutionary history.

**The System**

**Desert, Forgiveness and Grace in Evolutionary Perspective**

In the following three sections, I defend the following claims: First, emotions like anger, shame, and guilt explain our intuitions about the necessity of repayment. Second, these emotions evolved in part for self-protection. Third, one primary function for social systems of repayment (including norms regarding

---

[1] For a related discussion of virtue and desert, see Thomas Hurka, "The Common Structure of Virtue and Desert," *Ethics* 112.1 (2001): 6–31.

revenge, sacrifice, and blood money) is to constrain the negative effects of these emotions.

**Emotions Explain Repayment Intuitions**

Above, the focus has been on the phenomenon of repayment for past transgressions. Such repayment can take many forms, but perhaps the most paradigmatic variety of repayment is retributive punishment.[1] The very etymology of the word "retribution" reflects the concept of payback, as do philosophical and psychological understandings of retribution. In philosophy, retributive justifications for punishment are contrasted with utilitarian (or more aptly, consequentialist) justifications. Whereas utilitarian justifications tend to be *forward-looking*—appealing to the beneficial outcomes of punishment (e.g., deterrence, rehabilitation)—retributive justifications are *backward-looking*—appealing to the nature of the past offense and apportioning punishment to "fit the crime." In the context of punishment, the notion of fit is typically understood in connection with "just deserts": the punishment that transgressors *deserve* is justified in the sense that it "fits" their transgressions.

Importantly, philosophers who defend retributive justifications for punishment sometimes refer

---

[1] There may be a variety of distinct punishment phenomena with distinct underlying motivations. For instance, contempt and disgust may motivate punishments such as ostracism or exile. See e.g., Paul Rozin, Laura Lowery, and Jonathan Haidt, "The CAD Triad Hypothesis: A Mapping Between Three Moral Emotions (Contempt, Anger, Disgust) and Three Moral Codes (Community, Autonomy, Divinity)," *Journal of Personality and Social Psychology* 76.4 (1999): 574–86. For the sake of simplicity, I focus on retributive punishment, which is substantially linked to anger and guilt.

explicitly to emotions like anger and guilt.[1] Beliefs about the fittingness of punishment in response to past transgressions are justified in part because our feelings of anger and guilt testify to its fittingness. For instance, anger is sometimes a fitting response to injustice, and it motivates a retributive response to injustice. Likewise, guilt is sometimes a fitting response to one's own transgressions,[2] and it motivates guilty parties to accept punishment.[3]

In psychology, an extensive body of research suggests that the focus on just deserts is the main motivator for punishment. Though people often state that deterrence is their main concern in punishing, their decisions about punishment are better predicted by retributive considerations.[4] For instance, John Darley and Kevin Carlsmith have published a considerable body of evidence that when assigning punishments to hypothetical criminal offences, experimental participants attend almost exclusively to indicators of desert (e.g., the

---

[1] See e.g. Walter Berns, *For Capital Punishment: Crime and the Morality of the Death Penalty* (New York: Basic Books, 1979); Justin D'Arms and Daniel Jacobson, "Sentimentalism and Scientism," *Moral Psychology and Human Agency: Philosophical Essays on the Science of Ethics* (ed. Justin D'Arms and Daniel Jacobson; Oxford: Oxford University Press, 2014), 253–78.

[2] Moore, *Placing Blame*.

[3] Peter DeScioli and Robert Kurzban, "Mysteries of Morality," *Cognition* 112.2 (August 2009): 281–99.

[4] Jonathan Baron and Ilana Ritov, "Intuitions about Penalties and Compensation in the Context of Tort Law," *Journal of Risk and Uncertainty* 7.1 (1993): 17–33; Jonathan Baron and Ilana Ritov, "The Role of Probability of Detection in Judgments of Punishment," *SSRN Electronic Journal* 1.2 (2009): 553–90; Kevin M. Carlsmith, "On Justifying Punishment: The Discrepancy Between Words and Actions," *Social Justice Research* 21.2 (2008): 119–37.

seriousness of the crime) rather than deterrence. These results are robust even when participants are asked to focus exclusively on the deterrent effects of punishment. Moreover, Carlsmith and others find that the severity of punishments correlates with ratings of moral outrage at the given crime.[1] In effect, payback motives are deeply rooted in human psychology and they manifest themselves in decisions regarding punishment.

**Evolved Emotion and Self-Protection[2]**

There is a substantial body of research and theory suggesting that motives for giving and receiving payback are aimed at deterrence and include emotions of anger and guilt (among others). Consider first the claim that payback is an adaptation for deterrence. To think about the nature of payback motives, Robert Frank considers a situation like the following:

> Suppose Smith grows wheat and Jones raises cattle on adjacent plots of land. Jones is liable for whatever damage his steers do to Smith's wheat. He can prevent damage altogether by fencing

---

[1] John M. Darley, Kevin M. Carlsmith, and Paul H. Robinson, "Incapacitation and Just Deserts as Motives for Punishment," *Law and Human Behavior* 24.6 (2000): 659–83; Kevin M. Carlsmith and John M. Darley, "Psychological Aspects of Retributive Justice," *Advances in Experimental Social Psychology* 40 (2008): 194-236; Kevin M. Carlsmith, John M. Darley, and Paul H. Robinson, "Why Do We Punish? Deterrence and Just Deserts as Motives for Punishment," *Journal of Personality and Social Psychology* 83.2 (2002): 284–99; Kevin M. Carlsmith, "The Roles of Retribution and Utility in Determining Punishment," *Journal of Experimental Social Psychology* 42.4 (July 2006): 437–51.

[2] The ideas in this section owe a great deal to the work of David P. Barash and Judith Eve Lipton, *Payback: Why We Retaliate, Redirect Aggression, and Take Revenge* (Oxford: Oxford University Press, 2011).

> his land, which would cost him $200. If he leaves his land unfenced, his steers will eat $1000 worth of wheat. Jones knows, however, that if his steers do eat Smith's wheat, it will cost Smith $2000 to take him to court.... Smith threatens to sue Jones for damages if he does not fence his land. But if Jones believed Smith to be a rational, self-interested person, this threat is not credible. Once the wheat has been eaten, there is no longer any use for Smith to go to court. He would lose more than he recovered.[1]

One strategy to resolve Smith's dilemma is to convince Jones that he is a vengeful person and that he will take Jones to court even if he stands to lose money by doing so. The thought is that a motive for payback plays this very role by committing people to courses of action that are spiteful in the short run. Nevertheless, in the long run, this motive leads to a reputation for vengefulness that deters future offenses. So on this picture, payback motives functions to create a reputation for vengefulness that deters future transgressions.

While deterrence is the function of these strategies, deterrence cannot be the agent's immediate aim in punishing. This is because the short term cost of punishment will almost always eclipse the long term benefits of deterrence (in shortsighted organisms, at least), making organisms incapable of reaping those benefits. This is why the motivational structure of payback motives must be backward-looking in that they aim to punish *past* provocations in a way that is somewhat insensitive to immediate gains.

---

[1] Robert H. Frank, *Passions Within Reason: The Strategic Role of the Emotions* (New York: Norton, 1988).

Apparently, humans do have such a motive, and that motive is anger. Several lines of behavioral evidence to this effect come from research in behavioral economics, in which participants forgo monetary rewards in order to punish those who violate fairness norms. For instance, in public goods games, players will take on monetary costs to punish free riders (those who benefit from the investments of others without paying any cost themselves).[1] Moreover, punishers report feelings of anger at the free riders and they punish even when interactions are not repeated and even at the end of the last round of the game (after which no one will benefit from the punishment). Clearly, participants in these and other games place some value on payback, and these payback behaviors have been explicitly linked to anger and moral outrage.[2]

If we look even further back in our evolutionary history and further out in the animal kingdom, the dominant punishment strategies in animal societies are ones that that deter future transgressions by "teaching" transgressors to desist (though not necessarily via reputation).[3] Moreover, these punishment strategies

---

[1] Ernst Fehr and Simon Gächter, "Altruistic Punishment in Humans," *Nature* 415.6868 (January 10, 2002): 137–40; M. A. Nowak, K. M. Page, and K. Sigmund, "Fairness versus Reason in the Ultimatum Game," *Science* 289.5485 (September 8, 2000): 1773–75.

[2] Joydeep Srivastava and Francine Espinoza, "Coupling and Decoupling of Unfairness and Anger in Ultimatum Bargaining," *Journal of Behavioral Decision Making* 22 (2009): 475–89; Rob M. A. Nelissen and Marcel Zeelenberg, "Moral Emotions as Determinants of Third-Party Punishment: Anger, Guilt, and the Functions of Altruistic Sanctions," *Judgment and Decision Making* 4.7 (2009): 543–53.

[3] T. H. Clutton-Brock and G. A. Parker, "Punishment in Animal Societies," *Nature* 373.19 (1995): 209-16. Though I do not mean to imply that this is the only function of anger.

plausibly require a motive like anger to implement, for the very reasons given above (the long term benefits of punishment are eclipsed by short term costs).

Whereas anger motivates punishment of various kinds, evolutionary theorists suggest that emotions like shame and guilt attempt to mitigate punishment through appeasement. For instance, shame in humans appears to bear a close relationship to appeasement and concealment displays in other animals.[1] Likewise, guilt and perhaps conscience more generally may be adapted for mitigating the consequences of punishment by signaling sincerity of an apology or otherwise repairing one's reputation after a transgression.[2] Moreover, one of the most effective ways of appeasing a punisher, and of signaling sincerity of apology, is to communicate acceptance of punishment for the relevant transgression. That is, these evolutionary explanations appear to provide a clear explanation for why shame and guilt would lead one to believe in the appropriateness of retributive punishment as a response to one's own transgressions.

All this suggests that one central function of these emotions is self-protection. Whereas anger protects the self by coercing those who punish and their audiences to comply with one's wishes or to desist from

---

[1] See Dacher Keltner, "Signs of Appeasement: Evidence for the Distinct Displays of Embarrassment, Amusement, and Shame," *Journal of Personality and Social Psychology* 68.3 (1995): 441–54.

[2] See Cailin O'Connor, "Guilt, Games, and Evolution," *Emotion Researcher* (ed. Andrea Scarantino): http://emotionresearcher.com/guilt-games-and-evolution/ (accessed 4/11/17); DeScioli and Kurzban, "Mysteries of Morality;" Daniel Sznycer et al., "Regulatory Adaptations for Delivering Information: The Case of Confession," *Evolution and Human Behavior* 36.1 (2015): 44–51.

transgressions, guilt and shame protect the self by strategically manipulating those who punish. Moreover, it is their role in self-protection that generates intuitions about the appropriateness of retributive punishment.

**Systems of Repayment Constrain Evolved Emotions**

In a so-called "state of nature," emotions like anger obviously have the power to create reverberating patterns of revenge.[1] Consequently, in almost every culture, systems of norms have been constructed that function to constrain its effects.[2] Many of these systems begin with norms of proportionality and grant that the kin of a victim have the right to inflict proportional harm on a transgressor. Some systems transition to allow for blood money penalties as a substitute for bodily harm. Others eventually phase out the kin right altogether, in favor of institutional punishment overseen by feudal lords or authority figures.

Each of these changes clearly adds an additional level of constraint on revenge motives, so that each restriction can be expected to further diminish the level of vigilante justice and revenge in a given population. Thus, it seems clear that these systems function to limit the outgrowth of bloodshed and social disorder that results from revenge motives.[3] For instance, the decline

---

[1] The clearest cases of this occur in cultures of honor. See e.g. Richard E. Nisbett and Dov Cohen, *Culture of Honor: The Psychology of Violence in the South* (Boulder, CO: Westview Press, 1996).

[2] For a detailed defense of the ubiquity of revenge and the institutions that constrain it, see especially Martin Daly and Margo Wilson, *Homicide* (London: Transaction Publishers, 1988), chap. 10.

[3] Though it is an interesting question, it matters little whether that functionality is a product of intentional design or cultural evolution.

of kin rights in English law and the establishment of more institutional forms of retribution primarily benefitted feudal lords and kings in large part by maintaining order and limiting bloodshed.[1] Moreover, the role of norms regarding blood money is to provide some independent currency that can satisfy the motive to retaliate through substitution.

Old Testament practices of repayment and sacrifice are no exception to these patterns. First, Numbers 35:33 codifies a retributive norm concerning homicide: "...blood pollutes the land, and no expiation can be made for the land, for the blood that is shed in it, except by the blood of the one who shed it" (NRSV). Moreover, the *talion* ("An eye for an eye..." Lev 24:17-21) is best understood as a way of limiting the scope of the retributive norm via a norm of proportionality. Third, many of the rituals of sacrifice in the Torah function to substitute animal blood for the human blood that would otherwise be required as payment for transgressions against either God or one's fellow human.[2] Stott captures the basic elements of substitutionary animal sacrifice in this way:

> First, blood is the symbol of life.... Second, blood makes atonement, and the reason for its atoning significance is given in the repetition of the word "life." It is only because "the life of a creature is in the blood" that "it is the blood that

[1] Daly and Wilson, *Homicide,* chap. 10.

[2] For instance, one function of guilt offerings seems to be to make atonement for "less severe" offenses against other Israelites (such as those involving dishonesty) or offenses for which proportional compensation might be deemed inappropriate (e.g., having sex with a female slave).

makes atonement for one's life." Third, blood was given by God for this atoning purpose...[1]

I do not wish to overemphasize the role of sacrificial substitution in satisfying human revenge motives, for a large part of this sacrificial system is directed at atoning for various kinds of impurities and at man's relationship with God.[2] Accordingly, a greater part of the system functions to absolve feelings of guilt concerning transgressions against God or to absorb God's wrath against the Israelites.[3] This kind of

---

[1] Stott, *Cross of Christ*, 138.

[2] As Mary Douglas keenly notes (*Leviticus as Literature* [Oxford: Oxford University Press, 1999], 232) much of the sacrificial laws, especially in Leviticus, concern purity and food rather than vengeance, but she does not deny an implicit analogy between killing animals and homicide, which underpins the connection between animal sacrifice and payment for homicide: "Levine points out the wordplay by which Leviticus writes about shedding animal blood in terms that are usually used for homicide, but he stops short of interpreting the laws about animals as lower-key representations of homicide laws. There is no call to do so, since the language already serves well enough to dramatize the extreme gravity of the offence. Furthermore, Leviticus is about sacrifice and meat for food; the teaching on homicide is given elsewhere, in Genesis and in the Book of Numbers. The wider lessons are implicit. Genesis makes the reverse word-play, homicide described in terms of eating: God tells Cain that the ground... has opened its mouth to receive your brother's blood from your hand" (Gen 4:11).

For a study on the relationship between atonement-for-sin and atonement-for-impurity in the Torah, see Jay Sklar, *Sin, Impurity, Sacrifice, Atonement: The Priestly Conceptions* (Sheffield: Sheffield Phoenix Press, 2005). Sklar argues that the Hebrew verb translated as "atonement" (*kipper*) is univocal and involves a form of substitution in both uses. See esp. Parts I and IV.

[3] For a discussion of how God's wrath should be understood, see Wiegman, "Divine Retribution in Evolutionary Perspective." For an opposing view, see Oliver D. Crisp, "Divine Retribution: A Defence," *Sophia* 42.2 (October 2003): 35–52.

absolution or satisfaction is necessary because God is understood as the authority over the system of law, and as such all violations of the law code are ultimately thought of as transgressions against God.

To sum up, the Old Testament law functions in much the same way as systems of blood money and repayment in other cultures.[1] It constrains or limits revenge and in addition absolves feelings of guilt and shame. Importantly, blood is the primary currency of substitutionary repayment. From within this system, the sacrifice of blood or the transfer of money as a means of payback is seen as morally justified.

If Jesus' death is understood by analogy with this system, it is natural to understand it too as a justified moral transaction.[2] Nevertheless, when we reflect on the psychological and evolutionary origins of payback more generally, this moral justification seems doubtful. Systems of payback and punishment *appear* to provide moral permission for punishments within certain bounds, but our beliefs that punishment is *positively required* appear to derive from our evolved instincts for self-protection. Moreover, it is these latter motives that the system constrains and for which it offers substitutes

---

1 Though the law code clearly has other functions. For instance, purity norms also function to keep the Israelites "set apart" from surrounding nations, an aim which probably had instrumental value for God's plan of salvation.

2 Indeed, recent defenses of substitutionary atonement (Porter, "Rethinking the Logic of Penal Substitution;" Murphy, "Not Penal Substitution but Vicarious Punishment.") attempt to show how such substitution (or vicarious punishment) could be morally justified. On the view I articulate below, the crucifixion actually reveals that substitution is not a morally required transaction, and that the system of retributive punishment for transgression is itself morally bankrupt.

for physical harm to offenders. I think the overall effect of these considerations is to cast further doubt on the moral value of payback as well as the ultimate moral justification for the systems that license it.

Given that humans are in some sense stuck with vengeful motives as a result of our evolutionary history, we can understand why such a system might be morally justified as a way of preventing bad outcomes. Nevertheless, this preventative rationale does not provide any positive support for the claim that payback really is morally required as a matter of justice. If we think evolved motives for self-protection are at the heart of our beliefs about payback, then we should doubt that payback has any ultimate moral worth,[1] regardless of how well a system might function to constrain it. In the following section, I offer an alternative picture of the atonement; one that does not view it as a moral transaction within systems of payback. Instead, "Jesus' substitution" is better understood as an ingenious means for extricating us from those very systems and disabusing us of the retributive beliefs and motives they were built to constrain.

**An Alternative**
**Hacking the System, Ending Repayment**[2]

---

[1] I make this argument in greater detail elsewhere, see Isaac Wiegman, "The Evolution of Retribution: Intuitions Undermined," *Pacific Philosophical Quarterly* 96.2 (2015): 1-26.

[2] After writing the initial draft of this paper, I became aware of the work of Rene Girard and its application to atonement theory, as found in T. Scott Daniels, "Passing the Peace: Worship That Shapes Nonsubstitutionary Convictions," *Atonement and Violence: A Theological Conversation* (ed. John Sanders; Nashville: Abingdon, 2005), 125–48. My account of the atonement bears some structural

The natural place to start is with the bad news: we are stuck with a psychology that is shaped by evolution for self-protection. This is problematic for several reasons. First, self-protection is antithetical to love. Aside from that, our natural way of protecting ourselves creates cycles of revenge, barriers to trusting others, and even barriers to the kind of self-love that is necessary for self-giving love.

Consider first that the central action-tendencies of anger and guilt are toward self-protection. Certainly, mature, controlled, and directed anger *can* motivate a kind of confrontation that is essential for loving relationships. Nevertheless, we can see its more impulsive effects in young children and adolescents, who are much more likely to lash out or retaliate when angered. Moreover, when we lash out in anger, our aim (whether we realize it or not) is to guard ourselves by coercing others to comply with our present and future desires. But this coercion is clearly not out of love for the other, nor does it naturally induce love in its object. Likewise, mature guilt can motivate necessary reflection on one's failings, but as suggested above, its evolved effects may be aimed at a kind of manipulative appeasement. If this is right, then anger and guilt are Machiavellian in that they aim to maintain one's social standing, emotional well-being, and (ultimately)

---

similarities with this work, but my account does not depend on Girard's mimetic theory of desire. Moreover, I think my account also provides a better explanation of the connection between Jesus' crucifixion and the sacrificial tradition of the Old Testament more broadly, and not just the scapegoat ritual.

reproductive success by controlling others.[1] This kind of coercion and control is clearly antithetical to love.

Moreover, anger and guilt can maintain cycles of retaliation, mistrust, and self-doubt. In the case of anger, this is obvious even in more mundane interpersonal situations. When we are transgressed against, our anger and resentment readily leads to attributions of malicious intent that can be rekindled over time through rumination and record keeping.[2] When we transgress against others, our shame and guilt leads to self-punishment and doubts about our own goodness or worthiness to be loved. Once these doubts have taken root, they transfer to others. If I am rotten to my core, then others must be equally so.[3] Doubts are multiplied when we realize that the good will of others can be hostage to their indelible memory of our past transgressions against them. To sum up, the main problem is with trust. Forgiveness and reconciliation are impossible without trust, and the arms race of self-protection that anger and guilt initiate puts up thick barriers to trust.

It is inevitable that this lack of trust will extend to our relationship with God. We know that we have not treated God or creation with the kind of respect and care on which vibrant relationships depend. Moreover, we imagine quite naturally that God gets angry just like we

---

[1] See Paul E. Griffiths "Basic Emotions, Complex Emotions, Machiavellian Emotions," *Philosophy and the Emotions* (ed. Anthony Hatzimoysis; Cambridge: Cambridge University Press, 2003), 39–67.

[2] Brad J. Bushman et al., "Chewing on It Can Chew You Up: Effects of Rumination on Triggered Displaced Aggression," *Journal of Personality and Social Psychology* 88.6 (2005): 969–83.

[3] See e.g., Brene Brown, *Rising Strong: The Reckoning. The Rumble. The Revolution* (New York: Spiegel, 2015), chap. 6.

do. Out of our sense of guilt, we imagine that we must work hard to appease God. Out of our shame, we imagine that we must hide from God's displeasure.[1]

The trust required for reconciliation with others and with God simply cannot penetrate a heart so well-armored against rejection, retaliation, and betrayal. To see this, imagine that someone else's child died as a result of your own carelessness or indifference or malice. Moreover, suppose that this person wants to forgive you and even wants to allow you to go unpunished for your transgression. "I forgive you," they might say. "I will not be angry or resentful. I will not hold this against you. I will inflict no suffering, ask no favor, extract no penalty, require no blood money, all because I love you and you are worth loving. I will not play on your guilt or shame you into repentance. Certainly, I hope that you repent, but I also hope that you will forgive yourself and live free of guilt and shame. Though I loved my child dearly and am dumb-struck with pain at this loss, my love for you is without limit. I want you to live as well as my child might have lived." Could you possibly believe such a statement? Could the statement lead to full restoration of your relationship with the parent? Could you allow yourself to give and receive love from such a person in the ordinary course of life? Or would you be haunted by doubts about your own worthiness to be loved? Or doubts about their sincerity or resolve in keeping

---

[1] The story of Adam and Eve's fall from grace captures this dynamic vividly. A popular children's Bible story book encapsulates this part of the narrative in this way: "And a terrible lie came into the world. It would never leave. It would live on in every human heart, whispering to every one of God's children: 'God doesn't love me.'" Sally Lloyd-Jones, *The Jesus Story Book Bible* (Grand Rapids: Zondervan, 2007), 17.

resentment at bay? Would you feel that you are forever in this person's debt, party to a relationship with an insurmountable power imbalance? Might you end up forever working to try to undo the damage that you had done?

I suspect that no agent within "the system" can honestly make the statement that this hypothetical parent makes, and because of this, no one could possibly believe it. Given the massive incentives for self-protection within human psychology and the systems that constrain those incentives, it is obvious to everyone that no one could benefit materially from sending such a message. So for all intents and purposes, no one would ever take it seriously. The barriers to trust would be maintained even if some irrational, beleaguered or misguided parent (as evaluated by ordinary human standards) were to send such a message.

I believe that the beauty of the crucifixion is that it can credibly convey this exact message of self-giving love. The basic barrier to trust is the entire economy of self-protection constituted by our own self-protective emotions and by systems that codify the exchange of offense for offense and blood for blood. But the economy exists because each individual has limited resources with which to pursue their ends. That messages of forgiveness (like the parent's above) cannot be believed is a byproduct of scarcity.

At the cross, a horrible cost is extracted from an innocent son. But the son is a divine agent, a child of unimaginable "wealth," who is not permanently subject to scarcity and who therefore needs not participate in the economy of self-protection. Jesus' death is proof that he shares our humanity and proof also that the resulting harm and relational brokenness was real. Nevertheless, the resurrection is proof of the parent's great wealth and of the son's divinity. I said above that no agent within the system could credibly make the statement above. But if we come to believe that God and his son do not operate within the bounds of our system of self-protection, then this kind of forgiveness becomes credible. Though the cross is immense in its brutality, corruption, and brokenness, though it is among the worst crimes we might imagine ourselves committing, the good news is that God is willing and able to forgive it without punishment or penalty.

Moreover, the sacrifice and forgiveness of the Father and Son are out of love, with no ulterior motive. Thus the message of our own worthiness to be loved achieves a credibility that was previously unattainable. It becomes possible to forgive *ourselves* by letting go of our own shame and guilt. If this is right, then we can begin to let go of our motives for self-protection.

At this juncture, the promise that we will share in the resurrection aids in our escape. The resurrection is not only proof of divinity; it is also a promise that those who follow Jesus will share in the resurrection. We are adopted into a family of unimaginable wealth, and like Jesus, we can be restored from any insult or harm. Like him, we can give up on the economy of self-protection, together with its currency of debt and repayment. It becomes possible for us to make the same kind of sacrifice for others that Jesus has made for us. Moreover, we sacrifice for others not because we want to earn this grace, but out of gratitude for the grace that has already been extended.

On this view, it remains correct to say that the penalty for our transgressions falls on Jesus. After all, the Hebrew system of repayment was instituted by God, who has ultimate authority over all transfers within it. So it is God's prerogative to decide that Jesus' blood is substitute for the death that we "deserve." In this procedural sense, the consequences of our transgression within that system are transferred to Jesus. Nevertheless, we need not suppose that this transfer has any real moral weight, so long as the system is understood merely as a way of curbing the destructiveness of our evolved emotional responses. God is not thereby enforcing a moral requirement to collect a debt, since God's moral stake in the system may be merely to prevent the bad outcomes that accrue to human psychology (e.g., anger, guilt, etc.). Nevertheless, viewed from outside the system, the purpose of the transfer is not to satisfy a moral requirement but instead to meet the arbitrary demands of human psychology which shaped the system in the first place. Nevertheless, payment of this "debt" makes possible our exit from that system.

I believe that the elements of this escape plan can be found in in Jesus' enactment of communion. On the night he was betrayed, Jesus poured wine into a cup and called it his blood, asking his disciples to drink thereof. If blood is the currency of repayment for sins, then this act is like burning currency.[1] Why would someone burn currency? One reason would be to signify one's wholesale rejection of a monetary/economic system. If so, then a symbolic function of communion may very well be to say, in effect, "the currency of payback has ended for me, repayment for transgression is no more."

---

[1] It also symbolically violates prohibitions against drinking blood, which are justified by the fact that God is the owner of all blood and sanctions its use only within the sacrificial system "to make atonement."

Perhaps this is how the "new covenant of Jesus' blood" should be understood. Perhaps exit from the economy of self-protection is the kind of Exodus that the Passover meal of communion commemorates.

According to some, Jesus' enactment of communion suggests that his death must "be appropriated individually if its benefits... are to be enjoyed."[1] If one benefit of the cross is rescue from the system, then its efficacy requires the individual's wholesale rejection of the system. One cannot claim to reject an economic system wholesale if one actively manages an account within it.[2] This may explain why and in what sense the acceptance of Jesus death is necessary for salvation (here understood as escape from the system). Escape from the system means both forgoing repayment for transgressions against oneself and accepting that there will be no repayment for one's own transgressions (such as complicity in the crucifixion).[3] To demand repayment of others or to attempt repayment for oneself are both equally inconsistent with this commitment.[4]

---

[1] Stott, *The Cross of Christ*, 73.

[2] See especially the parable of the unforgiving debtor in Matt 18:21-35. Similarly, commitment to reject the system wholesale requires relinquishing one's positive assets within it. If one rejects the system and the debts one owes within it, one must also give up one's supposed earnings within it. In other words, it seems one cannot only give up on the negative desert that nagging guilt suggests to us without also giving up on the positive desert to which our soaring pride might lay claim. The parable of the workers in the vineyard in Matt 20:1-16 is clearly relevant here.

[3] For a beautiful and brief summary of traditional views on complicity, see Stott, *The Cross of Christ*, 63.

[4] Two tangential points are worth noticing here. First, the commitment to reject the system entails that one will forgo violence

This way of understanding the good news has many additional benefits. On this view, God need not be seen as endorsing any moral requirement of repayment or as transferring moral desert from one party to another or as participating in the moral punishment of an innocent. Moreover, there appear to be many elements of scripture and tradition to which it could be fruitfully applied.

**Conclusion**

To sum up, if we come to believe that our intuitions about payback are defective, then traditional understandings of the atonement need to be replaced. This alternative picture then is an interesting way of

---

for the sake of revenge or retributive punishment, but it does not entail total pacifism. One need not forgo violence that is instrumental for some morally valuable end, such as violence for the sake of protecting another. One who understands communion in the way I suggest may in good conscience use force to restrain a dangerous criminal or perhaps even participate in war to protect a people group from genocide. This kind of violence is clearly distinct from revenge and retribution, since its justification is forward-looking rather than backward-looking, as desert-based justification tends to be. I would even go so far as to say that ideal legal institutions might be justified in having "desert-based" policies, if those policies are the best at preventing harms via deterrence. The ultimate justification for such policies would be forward-looking, even though the individual justification for each act of punishment was backward-looking. See John Rawls, "Two Concepts of Rules," *The Philosophical Review* 64.1 (January 1955): 3. Second, I take it that an attempt at repayment for one's own sins is distinct from reparations. Reparations aim at restoring relationship or repairing harm, whereas repayment is fundamentally an act of self-protection. One *repays* in order to protect oneself from vengeance or hard feelings. The intended beneficiary of repayment is oneself, whereas the intended beneficiary of *repair* is someone else.

reformatting our understanding of the atonement to eliminate the moral legitimacy of payback and to reflect the emerging scientific picture of how evolution has shaped our moral outlook, often for the worse.

Of course, this is just a sketch of the good news that leaves many questions unanswered. For instance, what role does God the father play in the crucifixion? Can that role be justified from outside the system as well? How are we to understand scriptures that seem to endorse a moral commitment to the Old Testament system of repayment? And finally, on this view of the atonement, how is the grace given different from "cheap grace?"

The last question is particularly pressing. Outside a system of debt collection, God's forgiveness and grace looks to be completely free and without cost, as cheap as it could possibly be. Here, the tendency of many evangelicals is to insist that this is why sinners must be convinced that they truly deserve horrendous suffering as punishment for their sins. I think this misunderstands the meaning of "cheap grace." Cheap grace is not cheap because of the amount of punishment that ought otherwise to be exacted by the grace-giver. It is cheap because of the wrongdoer's low estimate of their wrongdoing and its effects. Sinners cheapen grace when they minimize the value of their relationships, the damage they have inflicted upon others, and their own culpability for that damage. Nevertheless, punishment is not the only way of signifying the heaviness of these moral weights. One can admit fully to these weights without supposing that they can or should be repaid in any currency.[1] Even where anger, resentment and payback are entirely off the table, one can entertain weighty concerns about one's relationships to those one has wronged: Can the damage be undone? Is she willing to trust me anew? Am I able to love him as he deserves? Can I avoid hurting them again? Will our relationship change in some way because of what I have done? Even if one has never participated in an economy of self-protection, restoration of any relationship remains infused with meaning and weight. Even if payback is not required, forgiveness is the prerogative of the wronged. Thus it cannot be taken for granted.

Regardless, if divine grace ends up being cheap on this picture, the implications of this grace are no less costly. It demands that we extend *unlimited* grace toward those still trapped in the system of payback (cf. Matt 18:21-22), and it is easy to see that this may cost us everything this world has to offer us. In the end, we may count this cost as a small thing (cf. Phil 3:8), but it is no small thing to undergo the radical shift in perspective that this picture requires.

---

[1] In fact, for even the most mundane of sins against another, I believe it would be crass to suppose that one could provide any payment for compensation.

## *When Neuroscientists Speak Religiously*

## *Navigating Neuroscientific Metaphysical Claims*

*Dean G. Blevins*

If theologians and ministry practitioners possessed any doubt that recent neuroscientific insights would impact pastoral practice, they should place those doubts to rest. The summer, 2014 edition of *Leadership Journal* dedicated a large part of writing to the intersection of ministry and brain research. The edition included the interesting subtitle: "Neuro Ministry: How Brain Science Informs Discipleship." In the journal, managing editor Drew Dyck offers a number of wise cautions often repeated in other research and popular articles. However, Dyck also asserts: "I believe wise leaders use all the tools at their disposal to more effectively pursue their callings—and brain science is a powerful tool."[1] At other times journalistic approaches may seem playful as with Mark Oestriecher's article, "Neurons-Shmeuron," in a recent *Youthworker Journal*.[2] However, Oestreicher's writing acknowledges that neuroscience content appears to be ushering in a fresh

[1] Drew Dyck, "Study the Brain Without Losing Your Soul," *Leadership Journal* 35.3 (Summer 2014): 5.

[2] Mark Oestricher, "Neuron-Shmeuron: Why Should We Care About Adolescent Brain Development?" *Youthworker Journal* 30.5 (May/June 2014): http://www.youthworker.com/neuron-hmeuron-why-should-we-care-about-adolescent-brain-development/ (accessed 4/11/17).

challenge for theologians and ministry practitioners, if only to challenge those in confessional traditions to articulate the strengths and limits of neuroscientific thought for the sake of discipleship in local congregations.

Fortunately, theologians, including ministry practitioners, do have resources within the field of neuro-education—the mind, brain, and education movement. A cursory review of this literature reveals several sound introductions to the field, integrating neuroscience with educational practice.[1] As practitioners adapt these texts to contemporary congregational settings, they will find practical insights that educators derived from basic principles of neuroscience. However theologically motivated practitioners may be surprised when neuroscientists move from these engagements with education to more detailed neuroscientific explorations of the world at large—particularly researchers that engage religion and religious experience. Often these neuroscientific investigations include not only metaphysical, but also outright assertions of a religious nature. In the face of these claims, how might theologians respond?

---

1 Mariale M. Hardiman, *Connecting Brain Research with Effective Teaching: The Brain-Targeted Teaching Model* (Lanham: Scarecrow Education, 2003); David A. Sousa, ed. *Mind, Brain, and Education: Neuroscience Implications for the Classroom* (Bloomington: Leading Edge, Solution Tree, 2010); Tracey Tokuhama-Espinosa, *Mind, Brain, and Education Science: A Comprehensive Guide to the New Brain-Based Teaching* (New York: Teachers College Press, 2010); Judy Willis, *Research Based Strategies to Ignite Student Learning: Insights from a Neurologist and Classroom Teacher* (Alexandria, VA: Association for Supervision & Curriculum Development, 2006).

Confessional theologians, and particularly ministry practitioners, may opt for an explicitly theological response, based on normative claims according to sacred texts or religious dogma. Christian educators, for instance, might respond based on their understanding of the gospel, guided by theologians with similar experiences engaging aspects of science and psychology. However, it may well serve confessional theologians to adopt a different approach, one that posits questions from a broader, religious studies framework first. Recognizing that much of the work translating neuroscientific claims relies on a largely interdisciplinary framework, religious studies affords an opportunity to raise questions often overlooked within confessional theology proper. Providing this larger framework allows theologians and practitioners to raise questions concerning each neuroscientist's definition of religion before asserting a normative critique. This approach affords theologians an opportunity to appreciate the work of neuroscientists who are willing to engage weighty questions without foreclosing conversations prematurely. Ultimately, theologians may accept, modify, or reject these metaphysical claims. However, they will do so based on a more nuanced view of the philosophical and religious views in neuroscientific literature.

This writing asserts that utilizing religious studies as a hermeneutical framework provides a theoretical understanding of neuroscientific meta-physical thinking. The thesis proceeds in the following manner: First, the writing includes two engagements, or "cases in point," that periodically illustrate the range of dialog resident within neuroscientific explanations of a larger reality beyond typical brain activity. The study surveys how neuroscientists' metaphysical inclinations often include implicit or explicit causal claims for the history of culture (Ramachandran), the nature of the self (Damásio and Beauregard), for the power of human religious experience (Newberg and McNamara), and the place of religion (Sacks, Brown and Jeeves). The writing then turns to explore how theology traditionally responds to these engagements, both through early encounters with neuroscience and within the broader view of theology proper. Finally, the writing invites theologians, and ministry practitioners normally familiar with working within confessional traditions into the world of religious studies. This journey demonstrates the interdisciplinary nature of both neuroscientific speculations as well as general approaches within religious studies that afford explanatory systems.

**Turning to the Metaphysical**

As a case in point, theologians and ministry practitioners might begin with *The Tell-Tale Brain*, what originally seems a fairly innocuous text by well known author, and TED presenter, V. S. Ramachandran.[1] Ramachandran, known for his work with phantom limb syndrome, provides an interesting introduction to the world of neuroscience, anchored in what he calls a "pervasive evolutionary perspective."[2] At one point Ramachandran introduces the role of mirror neurons in the shaping of social behavior.[3] However, he expands

---

[1] V. S. Ramachandran, *The Tell-Tale Brain: A Neuroscientist's Quest for What Makes Us Human* (New York: W. W. Norton and Sons, 2011).

[2] Ramachandran, *The Tell-Tale Brain*, xiv.

[3] Ramachandran, *The Tell-Tale Brain*, 119-35.

this neuroscientific discovery to posit that mirror neurons provide the basis of all social behavior, even civilization itself.[1] At this point theologians and practitioners familiar with Sigmund Freud's *Civilization and its Discontents* may begin to suspect a reductionist and materialist view of life and dismissal of religion.[2] Yet, a surprising statement occurs late in the last chapter when Ramachandran suggests that there is something "more" at work than merely mirror neurons:

> The real drive to understand the self, though, comes not from the need to develop treatments, but from a more deep-seated urge that we all share: the desire to understand ourselves.... We wander—to our peril—into metaphysics, but as human beings we cannot avoid doing so.[3]

In the face of a seemingly materialist view of human life, Ramachandran appears to move beyond complete materialism. At the end of the epilogue of the work, the neuroscientist reveals a philosophical view that suggests a burgeoning metaphysic, even as he opposes intelligent design.

> As a scientist, I am one with Darwin, Gould, Pinker and Dawkins. I have no patience with those who champion intelligent design, at least in the sense that most people would use that phrase. No one who has watched a woman in labor or a dying child in a leukemia ward could possibly believe that the world was custom crafted for our benefit. Yet as human beings we have to accept—with humility—that the question of ultimate origins will always remain

[1] Ramachandran, *The Tell-Tale Brain*, 132-35.

[2] Sigmund Freud, *Civilization and its Discontents* (Standard ed.; New York: W.W. Norton, 1989).

[3] Ramachandran, *The Tell-Tale Brain*, 291-92.

> with us, no matter how deeply we understand the brain or the cosmos that it creates.[1]

How might theologians navigate this modest metaphysical, almost religious, view? Ramachandran's "wonder" may be anchored in the mystery of atoms[2] or established earlier in the writing through a "faith" that historical innovation opened the door for mirror neurons to guide human advancement.[3] For all of his materialist, evolutionary conviction, Ramachandran appears to leave open a door for a deeper, metaphysical, understanding of the world around him.

V. S. Ramachandran represents a series of neuroscientists that seem to explore not only the nature of brain, but also the seat of consciousness and the self. Probably the best representative of this movement might be António Damásio.[4] Damásio posits that consciousness requires a dynamic view of the whole brain, beginning in the primal emotions of the brain stem, interacting with varying cognitive nodes of the outer portions of the brain. Damásio argues for an aesthetic view of this interaction, weaving a "symphonic" view of various parts of the brain operating without the need of a conductor, yet yielding consciousness.[5] After developing an intricate understanding of the biological underpinnings of consciousness, Damásio then closes his text with a larger view of how consciousness evolves into a sense of the self, culture, religion and art.[6] Again,

---

[1] Ramachandran, *The Tell-Tale Brain,* 293.

[2] Ramachandran, *The Tell-Tale Brain,* 292.

[3] Ramachandran, *The Tell-Tale Brain,* 134.

[4] Antonio Damásio, *Self Comes to Mind: Constructing the Conscious Brain* (New York: Pantheon Books, 2010).

[5] Damásio, *Self Comes to Mind,* 23-25.

[6] Damásio, *Self Comes to Mind,* 288-97

even Damásio, a strong advocate for evolutionary theory, posits a view of the world that requires a "socio-cultural homeostasis" reminiscent of a more basic search for homeostasis in the brain. [1] Yet, Damásio allows for the creation of myths and religion as a part of this larger balancing act between the individual's sense of self and larger cultural flows. Damásio, rather than restricting his treatise to the seat of consciousness, pushes outward to compose a comprehensive view of the self, culture, religion, and art—another demonstration of a neuroscientist's journey into the world of metaphysics.

**Other Engagements**

Ramachandran and Damásio may represent more nuanced views within a generally materialist understanding of reality. However, other neuroscientists take much more seriously the role of religious experience and religion in general in defining the humanity and the self. The range of these studies seems so vast that any simple survey may do a disservice to the work. Survey texts like Barbara Bradley Hagerty's *Fingerprints of God* may serve as a better introduction for ministry practitioners.[2] However, it might prove beneficial to see how certain "crosscurrents" within these studies may take theologians to new places for conversation within broad metaphysical claims. Within the general study of religious experience, two particular neuroscientists stand out: Patrick McNamara and Andrew Newberg.[3]

---

[1] Damásio, *Self Comes to Mind,* 292-93.

[2] Barbara Bradley Hagerty, *Fingerprints of God: The Search for the Science of Spirituality* (New York: Penguin Books, 2009).

[3] Patrick McNamara, *The Neuroscience of Religious Experience.* New York: Cambridge University Press, 2009); Andrew B. Newberg, *Principles of Neurotheology* (Ashgate Religion and Science Series;

However, while both neuroscientists explore religious experiences, their conclusions often reveal different methodological and metaphysical assumptions. McNamara offers a highly nuanced treatise of both neuroscience and religious experience, while acknowledging that his view remains governed by western, theistic, views of religion and self.[1] While neither dismissive of religion nor necessarily an adherent, McNamara does acknowledge a deep appreciation of the role of religion. Ultimately McNamara sees religion in service to the "Self" or executive function of the mind. He states, that insofar as religion is about individuals, it can be seen as an exquisitely attuned set of cultural practices that assists Selves in the process of creating new human cognitive powers and capacities."[2] McNamara then details research in religious experience in a manner that remains indebted to William James throughout his study.[3]

Andrew Newberg seems to come closer to understanding religious experience as a particular phenomenon deserving attention. Newberg launched a seemingly impressive project dedicated to the study of neuroscience and religion which he calls neurotheology,[4] setting forth "first principles" to govern comparative studies attending to both scientific insights and broad religious claims throughout the rest of the text.

---

Burlington: Ashgate, 2010); Andrew Newberg and Mark Robert Waldman, *Born to Believe: God, Science, and the Origin of Ordinary and Extraordinary Belief* (New York: Free Press, 2007).

[1] McNamara, *The Neuroscience of Religious Experience,* ix-x.

[2] McNamara, *The Neuroscience of Religious Experience,* 6.

[3] McNamara, *The Neuroscience of Religious Experience,* 14-15, 80-81.

[4] Newberg, *Born to Believe,* 45.

Newburg's approach appears quite ambitious, even positing the possibility that this approach might serve as a "megatheology" to bridge a number of religious traditions.[1] However, rather than remaining neutral, Newberg allows particular space for religious experience including phenomenal, spontaneous events like near death experiences (NDE), that may imply non-physical correlates for said experiences.[2] The comprehensiveness of Newberg's work also serves as a bit of an apologetic for the necessity of religious experience as an undergirding for neurotheology. In similar fashion, other neurologists have staked the presence of the "mind" as a definitive non-material state on similar phenomena.[3]

**Case in Point**
**Almost Heaven**

Beyond MacNamara's and Newberg's scientific and theological engagement with religious experience, theologians and practitioners encounter other theoretical constructions of the self anchored in religious overtones. A case in point builds upon the NDE phenomenon, but in a remarkable fashion.

One of the more remarkable accounts of NDEs resembles an account familiar to many Christians of a young boy's account of going to heaven.[4] However, this

---

[1] Newberg, *Born to Believe,* 64-66.

[2] Newberg, *Born to Believe,* 145-83.

[3] Mario Beauregard, *Brain Wars: The Scientific Battle Over the Existence of the Mind and the Proof That Will Change the Way We Live Our Lives* (New York: HarperOne, 2012), 157-82; see also Mario Beauregard and Denyse O'Leary, *The Spiritual Brain: A Neuroscientist's Case for the Existence of the Soul* (New York: HarperOne, 2007).

[4] Todd Burpo. *Heaven is for Real: A Little Boy's Astounding Story of His Trip to Heaven and Back* (Nashville: Thomas Nelson, 2011).

account emerges from an apparent NDE of neurosurgeon Eben Alexander.[1] Alexander's account includes his claim that his brain was clinically gone leaving him clinically dead, as Alexander says "my mind, my spirit—whatever you may choose to call the central part of me—was gone."[2] The remaining narrative includes Alexander's journey to the "Core" where he feels the fullness of Love.[3] Alexander proceeds to rearticulate his life from this supernatural experience, including reconciling science and spirituality, the ultimate purpose for writing the book.[4]

Eben Alexander's account created quite a reaction, reminiscent of Burpo's account ,yet carried the unique authority of a practicing, seemingly prestigious, neuroscientist in the center of the story. However, ongoing investigations of Alexander's account revealed controversial elements within the story, including some questions of Alexander's own professional career and his accounting of his clinical death. Instead *Esquire* journalist Luke Dittrich asserts Alexander was actually not "dead" but in a chemically induced coma, where hallucinations often occur.[5] Building on this revelation, other skeptics, like Michael Shermer, appeal to similar hallucinogenic NDE accounts documented by another neurologist, Oliver Sacks, the famous chronicler of neurological phenomenon, including his recent text titled

---

[1] Eben Alexander, *Proof of Heaven: A Neurosurgeon's Journey into the Afterlife* (New York: Simon & Schuster, 2012).

[2] Alexander, *Proof of Heaven,* 16.

[3] Alexander, *Proof of Heaven,* 70-71.

[4] Alexander, *Proof of Heaven,* 73.

[5] Luke Dittrich, "The Prophet," *Esquire* 150.4 (December 2012): 88-128. http://www.esquire.com/features/the-prophet (accessed 12/23/2016).

*Hallucinations.*[1] Dittrich's work reminds us that even neuroscientists may well be drawn into claims that, while supporting religious assumptions, merit careful examination.

**Theological Responses**

Confronted by these metaphysical expressions in neuroscience, ministers can turn to theology, particularly Christian theology, as a resource to respond to implicit, if not outright, theological claims. Ministers might begin by exploring the various theological aspects within neuroscience such as debates around divine causation and the nature of the soul.[2] Obviously there might be normative, *a priori*, guiding principles such as resistance to any evolutionary claims or rejection of anything other than a Cartesian view of mind/body division. These arguments have occurred at times in the past.[3] However, such a move might truncate any conversation with neuroscience based on such broad based, sweeping dismissals. Could there be other approaches? Three

---

[1] Michael Shermer, "Proof of Hallucination," *Scientific American* 308.4 (April 2013); http://www.michaelshermer.com/2013/04/proof-of-hallucination/ (accessed 4/11/17); Oliver Sacks, "Seeing God in the Third Millienium," *The Atlantic* (Dec. 12, 2012): http://www.theatlantic.com/health/archive/2012/12/seeing -god-in-the-third-millennium/266134/ (accessed 4/11/17); see also Oliver Sacks, *Hallucinations* (New York: Knopf, 2012).

[2] Nancey Murphy, *Bodies and Souls, or Spirited Bodies?* (New York: Cambridge University Press, 2006); Robert J. Russell, Nancey Murphy, Theo C. Mayering and Michael A. Arbib, eds. *Neuroscience and the Person: Scientific Perspectives on Divine Action*, (Vatican City: Vatican Observatory along with the Center for Theology and the Natural Sciences, Berkeley California, 1999).

[3] Joel. B. Green, ed., *In Search of the Soul: Four Views of the Mind-Body Problem* (Eugene: Wipf and Stock, 2005).

theological frameworks have guided recent conversations between theology and science (or social science).

**Radical Orthodoxy**
**Violence or Peace**

Guided by the work of John Milbank, Catherine Pickstock, Graham Ward and others, the Radical Orthodoxy movement has both critical and constructive components.[1] Distrusting of any sacred/secular split, the movement argues that all of reality finds its "purchase" in God and seeks to demonstrate the limits of any philosophical approach that locates itself within secularism. Critically, Milbank argues that certain movements in postmodernity (particularly the postmodern deconstructionists that Milbank calls secular postmodernists) exist as extensions of the modern emphasis on secularity and violence as the ontology of nature.[2] Milbank believes that these movements assume that life at its deepest reality is violent. Milbank will concede that certain aspects of life occur due to the possibility of violence. Milbank, however, critiques the limits of most social theorists who implicitly posit that the purpose of life (metaphysically) remains at best a negotiation or survival in the face of ontological violence.[3]

---

[1] John Milbank, Catherine Pickstock, and Graham Ward, eds., *Radical Orthodoxy: A New Theology* (London: Routledge, 1999).

[2] John Milbank, *Theology and Social Theory: Beyond Secular Reason* (2nd ed.; New York: Wiley and Sons, 2006).

[3] Henry Spaulding, II, "Good Conscience or Good Confidence: A Postmodern Re-Thinking of Ethical Reflection in the Wesleyan/Holiness Tradition," *Wesleyan Theological Journal* 35.1 (Spring 2000): 41-66.

Constructively, Milbank and others would instead see the world participating with the divine. These theorists wish to rediscover certain pre-modern themes they feel better articulate reality in its participation with God in a path toward Augustine's "heavenly city" marked by peace and harmony. To emulate this path is the goal of theology. Theology represents not only a specific branch of human thought (alongside sociology, psychology, philosophy, etc.), it becomes the articulation of transcendent aspirations (and in some cases the apophatic contemplation of the limit of articulation) of all rationality. Persons are called to participate with God, using action to embody (incarnate) the heavenly city and thus reveal a peaceful ontology that is one with God. Participation, often described in liturgical form, becomes a term that unites aesthetic, transcendent, aspirations with political action. The result is a new community engendered in the reality of God.[1]

Admittedly, the Radical Orthodox movement has developed its notoriety in its more accomplished critical assessment of the limits of secular reason. Theologians and practitioners appropriating this view will appreciate the critique of a world encapsulated in violence. Such a theological framework might take some neuroscience speculation and recast it in light of God's gracious and general movement to reconciliation and peace.

**James Loder**
**Relationship with the Spirit**

The work of the deceased James E. Loder may provide a second theological framework to guide

[1] Catherine Pickstock, *After Writing: On the Liturgical Consummation of Philosophy* (Oxford: Blackwell Publishers, 1998).

conversations with neuroscientists. Loder himself has posited a neuroscientific view for his focus on "transformational moments" based on maintaining an equilibrium within the hemispheric portions of the brain.[1] However, Loder's larger perspective of the work of the Holy Spirit provides a different beginning point.

Loder originally posited a relational "logic" between human experience and our encounter with the Spirit of God.[2] Drawing from existential frameworks similar to those of Søren Kierkegaard, Loder crafts a model that describes what he calls the irreducibility of relationality between human and divine, a relationality that overcomes traditional duality. Loder and physicist William Neidhardt assert that this "logic" of transformation surfaces through the relational interaction of the human spirit with the Holy Spirit that he believes remains irreducible.[3] Loder believes he can best represent his model of relationality via the image of the Möbius Strip, a "strange loop" of self-relationality that may prove asymmetrical in relationship reflects a paradoxical quality in mathematics, music, and quantum physics. Loder argues that this structure provides a means for describing "irreducible relationality" between divine and human agency. Loder and Neidhardt write:

> In general, the model presents the asymmetric bipolarity of relationality, suggesting its inherent unity. The apparent two sides or edge

---

[1] James E. Loder, *The Logic of the Spirit: Human Development in Theological Perspective* (San Francisco: Jossey-Bass 1998), 36-37.

[2] James. E. Loder, *The Transforming Moment* (2nd ed.; Colorado Springs: Helmers and Howard, 1989).

[3] James. E. Loder and William J. Neidhardt, *The Knight's Move: The Relational Logic of the Spirit in Theology and Science* (Colorado Springs: Helmers and Howard, 1992), 21-53.

> of the Mobius band represent the two poles in a dynamic interrelatedness which via a 180° twist brings the apparent duality into a paradoxical unity. This relationality may actually represent an emergent property anchored in relational processes that extend beyond traditional physicalist or dualist categories.[1]

Loder's attempt to bridge between the work of the Holy Spirit and human experience provides another approach that affords theologians and ministry practitioners with a new understanding of how certain neuroscientific claims might well reflect not only observations on human experience but also reflections of the work of God's Holy Spirit in and through human creation.

**Don Browning**
**Correlating Ethics**

A third approach surfaces through the practical theology of Don Browning, who employs a "critical correlational" approach to establishing a dialog between theology and psychological theories.[2] However, Browning's method rests upon different criteria than either the peacefulness of Milbank or the relationality of Loder. Browning instead focuses on the moral psychological assumptions or ethical implications within any psychological theory, asking how implicit views of God, self and community might improve or hamper the overall well-being of persons and society.[3]

---

[1] Loder and Neidhardt, 55.

[2] Don Browning, *Religious Thought and the Modern Psychologies* (2nd ed.; Minneapolis: Augsburg Fortress, 2000).

[3] Browning, *Religious Thought,* 5-15.

Browning's view does appear to appeal to a "third category" (moral philosophy or ethics) as the arena for dialog. While such a move may seem to give up normative theological claims in favor of a mutually critical correlation between theology and psychology, Browning's own engagement with obligatory ethics seems to belie that concern.[1]

Browning's turn toward the ethical creates a middle ground for dialog between neuroscience values and questions of appropriateness and being, moving closer to a general metaphysic of "the common good" that may prove attractive to neuroscientists like Martha Farah and others who seek to understand the ethical nature of neuroscientific principles and practices.[2]

### Theology and Norms
### Moving Toward a Playful Alternative

Ultimately all three theological movements still accept certain theological criteria as *a priori* and normative for the conversation. As such, while these theories afford a means of assessing metaphysical claims, they may also foreclose others, particularly when alternative definitions of reality or religion occur. Could there be a more open, playful, range of theories that might at least bracket theological norms to open new lines of investigation and conversation among theologians, ministry practitioners, and neuroscientists? Religious studies, as a discipline and tradition, may hold the answer.

---

1 Browning, *Religious Thought*, 18-20.

2 Martha J. Farah, ed. *Neuroethics: An Introduction with Readings* (Cambridge: MIT Press, 2010).

## Religious Studies
## An Alternative Consideration

The previous theological approaches might provide an informative hermeneutic for understanding neuroscientific metaphysical claims. Still, each approach risks limiting the conversation due to *a priori* claims indebted to a theological reading. Perhaps a better approach for engaging a number of neuroscientific speculations rests within the world of religious studies. Religious studies represents a multi-disciplinary "movement" that employs diverse theoretical approaches in the study of religion, either in comparing religious movements or in positing a general theory of religion. Walter Capps notes that the discipline of religious studies, as a whole, entails both the task of discovery and also the task of articulating the intelligibility of religion.[1] Charles Jones argues that religious studies theories often draw from differing subdisciplines in the social sciences such as sociology, economics, psychology, cultural anthropology, and even a phenomenological view of religion as a subcategory in itself.[2] As such, the religious studies movement includes the more traditional category of psychology of religion, one that has often governed the dialog between psychology and religion.

## Psychology of Religion or the Religious Study of Neuroscientific Claims

---

[1] Walter H. Capps, *Religious Studies: The Making of a Discipline* (Minneapolis: Fortress Press, 2000), xiv.

[2] Charles B. Jones, *Introduction to the Study of Religion* (Great Courses Audio Series; Chantilly: The Teaching Company, 2007).

Traditionally, the field of the psychology of religion proved useful as a disciplinary field followed in lieu of the larger traditions that govern religious studies. The study of the psychology of religion includes a rich array of disciplinary conversation partners in fashioning an understanding of the role of religion and religious experience within persons and among communities. H. Newton Maloney notes the psychology of religion might be organized in various subthemes, at times guided by therapeutic concerns rather than educational interventions.[1] For theologians and ministry practitioners the largest categories might best be described as 1) Foundational Issues in the Psychology and Religion Dialog; 2) Personal Religious Experience: Spiritual, Existential, and Developmental Considerations; 3) Disciplinary Studies in Psychology and Theology; 4) Social Psychology and the Nature of Religion; and, 5) Applying Psychology and Religion.[2]

For all of the benefits of approaching metaphysical assertions within psychology of religion, the discipline may lack a sufficient explanatory system to understand all metaphysical and religious neuroscientific claims. As Daniel Pals notes, only Freud (and perhaps Jung) represent one of eight larger theories of religion established since the beginning of this discipline in the eighteenth century.[3] Other primary theories emerged through the work of anthropologists E. B.

---

[1] H. Newton Malony, ed., *Psychology of Religion: Personalities, Problems, Possibilities* (Grand Rapids: Baker, 1991).

[2] Dean G. Blevins, "Psychology of Religion," *Encyclopedia of Christian Education* (ed. George Thomas Kurian and Mark A. Lamport; Lanham: Rowman and Littlefield, 2015), 1007-08.

[3] Daniel L. Pals, *Eight Theories of Religion* (2nd ed.; New York: Oxford University Press, 2006), 12.

Tylor, J. G. Frazier and later E.E. Evans-Pritchard. In addition, major theories of religion surfaced from the writings of sociologist Emile Durkheim, economists Karl Marx and Max Weber, cultural anthropologist Clifford Geertz, and religionist Mircea Eliade. The broad depth of disciplinary resources within these fields expands the overall range of interpretive categories when exploring how neuroscientists use the term "religion" or "culture" in their metaphysical speculations.

For instance, Pals notes that the term "religion" may take on either substantive (or normative) definitions or more functional definitions.[1] A functional definition often situates the understanding of religion within a specific context, where the definitional categories might be "bounded" by the dialog. So, a conversation between neuroscience and Christianity might be bound by functional understandings of Christian belief and practice. Yet often neuroscientists make claims about "religion" that seem to be more broad and substantive, as a kind of commonsense definition of religion for everywhere and all time. However, these definitions often demand greater specificity since even belief in God might be too limiting for some religions (such as Buddhism). In this case Pals and others argue for a more general "sense of the sacred," a term that might prove actually inclusive of certain metaphysical claims within neuroscientific fields. It is important to note that the field of religious studies often pushes a greater sense of definition than many neuroscientists use when discussing religion, or even religious experience. Religious studies scholars dedicate entire texts to definitional concerns. Mark C. Taylor notes that the

[1] Pals, *Eight Theories of Religion*, 12-15.

multidisciplinary—and multicultural—engagement of comparative religions raises methodological and theoretical issues.[1] Theorists need to approach the dialog with care between neuroscience and religion (or metaphysics) with care.

Such diverse, interdisciplinary engagement proves reminiscent of the work of neuroscientists as they move from the laboratory to the conversation. Volney Gay notes that neuroscientists often employ a large array of disciplines in constructing their explanations of the workings of the brain.[2] Often neuroscientists draw from the fields of physics, chemistry, animal behavior, as well as social sciences like psychology and sociology, when postulating mental activity. At times neuroscientists make comparative "leaps" by associating brain behavior with the behavior of insects or other animals.[3] The range of multidisciplinary fields already informing neuro-scientific investigations seems to invite a similar dialog with a larger, more complex study of religion than often assumed.

**Case in Point**
**Religion as Baseball**

As a case in point one might turn to the work of neuroscientists and committed theologians Malcolm Jeeves and Warren Brown (2009) and their general

---

[1] Mark C Taylor, "Introduction," *Critical Terms for Religious Studies* (ed. Mark C. Taylor; Chicago: University of Chicago Press, 1998), 13-14.

[2] Volney P. Gay, "Introduction," *Neuroscience and Religion: Brain, Minds, Self, and Soul* (ed. Volney P. Gay; Guilford: Lexington Books, 2009), 12-14.

[3] Gay, "Introduction," 5.

depiction of religion as baseball.[1] Brown first adopted this metaphor out of his concern about the special use of "religious experience" as a particular category.[2] Brown, summarizing a number of neurological studies of religion, raises questions about whether religion reflects more a natural cognitive ability like music or whether religion more resembles a social cultural phenomena, like baseball.[3] In Jeeves' and Brown's summary overview of neuroscience, psychology and religion, the neuroscientists pick up this argument as a primary critique of finding a neurological basis for religious experience.[4] They write:

> With baseball as a conceptual model for religion, the neurological study of religion changes its approach. First, we would not expect to find a specific neurology of baseball—that is, no unique neurological systems that would contribute specifically to baseball and not to other forms of life. Baseball is neither sufficiently unitary as an experience nor sufficiently embodied in biology to study at the level of neurology. Second, we would not expect to find a neurological disorder specific to baseball, although such a disorder in a person might alter the participation and appreciation of the sport. Third, it would be somewhat far-

---

[1] Malcom Jeeves and Warren. S. Brown, *Neuroscience, Psychology, and Religion: Illusions, Delusions, and Realities about Human Nature* (West Conshohocken: Templeton Foundation Press, 2009).

[2] Warren S. Brown, "The Brain, Religion, and Baseball: Comments on the Potential of a Neurology of Religion and Religious Experience," *Where God and Science Meet* (3 vols.; ed. Patrick McNamara; Westport: Praeger Publishers, 2006), 2:229-44,

[3] Brown, "The Brain, Religion, and Baseball," 230-31

[4] Jeeves and Brown, *Neuroscience, Psychology, and Religion*, 99-101.

> fetched to imagine the evolution of the specific capacity for baseball or to argue for the survival advantages of baseball to individuals or social groups or to argue that the specific capacity for baseball is "hard wired." Rather, baseball is a complex social phenomenon. The reality of baseball "emerges" as it piggy-backs its activities and experiences—cognitively, neurologically, and evolutionarily—onto a large number of more general cognitive capacities and skills.[1]

What the authors leave unsaid in this social depiction of religion as baseball opens the door to multiple interpretations, depending on the dialog partner within religious studies. For instance, religion as baseball resembles the work of Durkehim, who basically posited that religion serves to "sacralize" the needs or values of society at large.[2] This view, while reductionistic, actually serves the metaphor well, since often sports carry within them specific religious connotations fueled by the social fan base (be it a college sport or professional sports team). However, religion as baseball might actually focus more on the "grammar," or thick description, within the sport, which approximates more closely the cultural perspective of Clifford Geertz who employs ethos and worldview together to create a descriptive rather than reductionistic view of religion.[3] Jeeves and Brown do not provide an answer to mediate between the two views.

Readers must understand that the following descriptive engagement does not serve to critique Jeeves

---

[1] Jeeves and Brown, *Neuroscience, Psychology, and Religion,* 100.

[2] Pals, *Eight Theories of Religion,* 107.

[3] Pals, *Eight Theories of Religion,* 281-89.

and Brown. Both theorists stand within a confessionally oriented theological tradition, underscored by theological and biblical frameworks that support their position.[1] However, their use of a more neutral approach to defining religion affords an opportunity to see how religious studies open the door to multiple perspectives without demeaning the work offered.

Admittedly, Jeeves and Brown actually come quite close to other recent changes in religious studies. Ann Taves notes that many leaders in the early study of religious experience often flattened the field based on core liberal Protestant perspectives that basically reduced the study of religious experience to implicit theological traditions that, in turn, defined that experience.[2] Neuroscientists often fall into this mistaken assumption of one, universal, generic view of religious experience. Only recently has the study of religious experience shifted to a new, more multivalent view of "experiences that are religious."[3] In other words, theorists are beginning again to examine personal accounts of religious insight from a wide array of social and religious perspectives. Taves notes that this shift opens the field of study, but also raises fresh challenges for researchers who rarely possess the requisite skills to transverse the varied fields of psychology, neuroscience, and religious studies.

In this case, Jeeves and Brown's depiction of religion as baseball opens the door for multiple

---

[1] Joel B. Green, *Body, Soul, and Human Life: The Nature of Humanity in the Bible* (Grand Rapids: Baker Academic, 2008).

[2] Anne Taves, *Religious Experience Reconsidered: A Building-Block Approach to the Study of Religion and Other Special Things* (Princeton: Princeton University Press, 2010), 4-5, 21-22.

[3] Taves, *Religious Experience Reconsidered,* 8-9, 16-55.

discussions on the nature of religion and religious experience that might help researchers and practitioners ask if either term (religion or religious experience) can be used in a universal sense or should actually be used to explore specific, contextual expressions. Rather than merely critiquing Jeeves and Brown's depiction, a religious studies exploration helps to open up both sides of the conversation around the nature of religion and experience within neuroscience.

**Principles for Conversation**

Religious studies, as a discipline, continues to move away from eighteenth century notions of "religion" as a substantive or universal description.[1] Instead, as religious studies explore other notions of the sacred, the discipline may provide comparative, functional categories to implicit metaphysical claims within neuroscience, whether around ideas of experience, culture, or "mindfulness."[2] The deeply interdisciplinary nature of both fields allows for a type of playful engagement while bracketing *a priori*, normative assumptions within theology proper.

Perhaps theologians might follow the work of Douglas John Hall in his work, *The Steward*.[3] While unpacking the biblical idea of the steward, Hall proposes several principles or "middle axioms" that worked as

---

[1] Robert A. Orsi, "Introduction," *The Cambridge Companion to Religious Studies* (ed. Robert A. Orsi; New York: Cambridge University Press, 2012), 1-15; Eric. J. Sharpe, *Comparative Religion: A History* (LaSalle: Open Court, 1986).

[2] Daniel J. Siegle, *The Mindful Brain: Reflection and Attunement in the Cultivation of Well-Being* (New York: W. W. Norton, 2007).

[3] Douglas John Hall, *The Steward: A Biblical Symbol Come of Age* (Grand Rapids: Eerdmans, 1990).

conversation partners with larger social concerns.[1] In like fashion, religious studies offers several operative principles, shaped in the form of questions that might help shape a conversation with neuroscience when positing a metaphysical/religious posture:

> Does the theorist use "religion" or "reality" within a given context or as a universal definition?
>
> Does "religion" (or whatever ultimate reality portrayed) include a one or a two tiered universe (i.e., the physical and the spiritual or just one of the two dimensions)?
>
> Does "religion/reality" appear to come from "above" (a sense of an in-breaking sacred), "within" (a sense of personal experience) or "among" (a sense of the social or communal)?
>
> Can the "religion/reality" portrayed be reduced to simple principles (i.e., reductionism) or does it display a deep sense of complexity?
>
> How does the "religion/reality" portrayed serve those who name and participate in its reality?

These questions provide a beginning point for understanding any metaphysical or any religious claim and build at least points of contact within confessional theologies that should also respond similarly.

**Conclusion**

Ultimately, as Justin Barrett notes, neuroscience may better serve as a conversation partner "within"

---

[1] Hall, *The Steward,* 123-25.

confessional theological traditions, since claims to a universal natural theology (which Newberg might favor) seem less tenable.[1] However, since many neuroscientists—as metaphysicians—seem to lie outside confessional theology, it appears that the use of "middle" principles and attentiveness to varying approaches to religious studies afford theologians and ministry practitioners space to appreciate specific questions of meaning and hope. This approach also allows theologians to selectively allow the claims within neuroscience to function as conversations partner without interfering with normative theological claims. The dialog, for neuroscientists and theologians alike, may prove quite fruitful.

---

[1] Justin L. Barrett, *Cognitive Science, Religion and Theology: From Human Minds to Divine Minds* (Templeton Science and Religion Series; West Conshohocken: Templeton Press, 2011), 148-67.

## *A Thoughtful Technophobia*

*Joshua Kira*

**Introduction**

The title "A Thoughtful Technophobia" was not intended as a way of demeaning other manners in which individuals have concerned themselves with the philosophical questions surrounding technology,[1] as if the present author has suddenly come to a conclusion that is finally well-thought. Instead, the title was meant to indicate the difference in the source of technophobia, drawing a distinction between a contemplative and speculative caution with regard to technology. A speculative form would be of the type that would attempt to predict the possibilities of technology in the future and the types of dangers, physically and otherwise, inherent to these possibilities. A typical example would be the recent fears from individuals as diverse as Stephen Hawking[2] and Elon Musk[3] concerning artificial intelligence. A contemplative, or thoughtful, technophobia would be a concern with technology that is born out of the present manner in

---

[1] For summary of the development of Heidegger's view on the relationship between science and technology, see Lin Ma and Jaap van Brakel, "Heidegger's Thinking on the 'Same' of Science and Technology," *Continental Philosophy Review* 47.1 (2014): 19-43.

[2] Rory Cellan-Jones, *BBC News*, http://www.bbc.com/news/technology-30290540 (accessed March 10, 2016).

[3] Tess Townshend, *Inc.com*, http://www.inc.com/tess-townsend/elon-musk-open-ai-safe.html (accessed March 10, 2016).

which individuals relate to technology. In line with this type of thinking would be concerns with the dehumanizing nature of technology, the effects of technology on social structures and stability, etc. This chapter will begin to delve into a thoughtful technophobia in interaction with the work of Martin Heidegger, whose particular brand of phenomenology attempted to examine technology as it stood in contemporary society and the way in which it could be a window into Being.

Martin Heidegger did not begin nor end the strain of technophobia that ran through 20th century phenomenological thinking. Worries concerning technology can be seen in both his predecessor and successor, Edmund Husserl[1] and Hans Georg Gadamer.[2] However, due to the influence of his rural upbringing and constant retreat to the Black Forest of Germany, he is, perhaps, the most pointed in his criticism. Heidegger is often seen as unusual in that he put a premium on human freedom and heralded the dangers of technology, and yet supported the coercive activity of the Nazi party and the technological advances of the German war machine. Yet, his support of those should not be immediately interpreted as a contradiction between theory and practice, but instead should drive a recognition, at least in the latter, that technology was not

---

[1] See Edmund Husserl, *The Crisis of European Sciences and Transcendental Phenomenology* (trans. David Carr; Evanston, IL: Northwestern University Press, 1970).

[2] See Hans-Georg Gadamer, *Truth and Method* (2nd ed.; trans. Joel Weinsheimer and Donald G. Marshall; New York: Continuum, 1989).

inherently evil from his perspective.[1] Instead technology displayed a mode of being within humanity that was both illuminating and obscuring, with his fear deriving from the inability to easily extract the prior activity from the latter. In looking at the manner in which Heidegger analyzes language and the technological life, it will be argued that this obscuring is a particular worry, especially to those of a Christian commitment.

**Heidegger and Technology**

When examining technology, Heidegger wants to refrain from approaching it as a subject that can be separated from its relationship to humanity. He argues strongly against this type of thinking in *Being and Time*,[2] where he claims that the being of *Dasein*, which is any being that is concerned about its Being, is Being-in-the-world (*in-der-Welt-Sein*).[3] The inseparability of Being and worldliness is such that there is a false dichotomy when attempting to see humans as subjects and the world as their object. It is this belief that leads him to question some of the manners and methods of an early 20th century view of science. Moreover, the close relationship

---

[1] There are times where Heidegger's view of technology is seen as almost solely negative. This can be due to a moralizing of the idea of *authenticity* in *Being and Time*, an ethical connection that is absent from his early ontology. For an example of this see Andrew Komasinski, "Anti-Climacus's Pre-emptive Critique of Heidegger's 'Question Concerning Technology,'" *International Philosophical Quarterly* 54.3 (September 2014): 265-277.

[2] Martin Heidegger, *Being and Time* (trans. Joan Stambaugh; Albany: State University of New York Press, 1996). Martin Heidegger, *Sein und Zeit* (Tübingen: Max Niemeyer Verlag, 1956).

[3] Heidegger, *Being and Time*, 53-62. The hyphenation of the original text is usually recognized as Heidegger's linguistic signifier of inseparability.

of Being and the world means that to understand Being, especially human being, is to examine it through the everyday activities and choices of humans. Thus, in a manner similar to his analysis of tool use in *Being and Time*, Heidegger attempts to understand technology through its relationship with humans in their everydayness (*Alltäglichkeit*).[1] Technology is, therefore, not a question of objective presence, but the product of a prior technological mode of life. In this mode of life, the essence[2] of technology can be found. In other words, a technological mode of being leads to technology and not the other way around.

To understand the technological life, Heidegger begins with the common understanding of technology as being instrumental. Technology is always teleological and therefore used, like a tool, for humans to enact their Being. He then, since a purpose is always present in the use of technology, sees it as being an issue of causality. The technological mode of life is a causal form of life that is oriented towards human goals. Heidegger begins to reframe the understanding of causing, especially as seen in Aristotle's account, to distill what commonly occurs in different forms of causation. Material, formal, final, and efficient causes are understood to be that which is responsible for bringing forth something, such as an

---

[1] Heidegger, *Being and Time*, 63ff. Heidegger's existential concerns can be seen in his claim, "The relationship will be free if it opens our human existence to the essence of technology." Martin Heidegger, "The Question concerning Technology," *The Question concerning Technology and Other Essays* (trans. William Lovitt; New York: Garland Publishing, 1977), 3-35, here 3. Available online at: http://www2.hawaii.edu/~freeman/courses/phil394/The%20Question%20Concerning%20Technology.pdf.

[2] See Heidegger, "Question concerning Technology," 3-5.

artifact or a piece of technology.[1] Yet, this responsibility is not independent from already existent possibilities. For example, to use Heidegger's illustration,[2] when a silversmith makes a sacrificial chalice to be used in religious rites, he is, in part, responsible for the appearance[3] of the chalice in its relationship to human life. This appearance, though, is not the creation of possibilities of enactment, since the chalice was always a possibility in the context of the humanly lived religious ritual. For this reason, Heidegger is careful to speak of the smith's activity in terms of "co-responsibility" (*mitschuld*) so that the context of existent possibilities is emphasized. However, without the silversmith, not only would the possibility of a lump of silver being part of religious activity remain hidden, but so also would those aspects of the religious life with which the chalice is related. By making the silver cup, the artisan shows what was already there, a type of human enactment, the religious life, in which a sacrificial chalice holds significance. Consequently, this bringing-forth (i.e., *poiesis*),[4] is foundational for the understanding of the technological life in that the possibilities for human enactment, which are hidden in nonuse, are then clarified by the bringing-forth of technology. Simply put, to use technology is to facilitate an appearance of enactment that shows humans their possibilities.

---

[1] Heidegger, "Question concerning Technology," 7.

[2] Heidegger, "Question concerning Technology," 6ff.

[3] This is why Heidegger connects *poiesis* with *apophainesthai* ("appear forth"), which helps to relate his work in "Question concerning Technology" with his understanding of revelation in *Being and Time*. See Heidegger, "Question concerning Technology," 8.

[4] Heidegger, "Question concerning Technology," 10ff.

In this way, Heidegger connects technology to truth and revelation. Consistent with his previous attacks of correspondence theories and continuing his phenomenological methodology, Heidegger sees truth (i.e., *allētheia*) as the unveiledness of what is previously concealed,[1] with truth understood as the quality of being revealed. Heidegger, in seeing technology as bringing-forth, connects that coming with revelation, such that technology's purpose is revelatory. Consequently, the technological life reveals something about the possibilities of human life. Yet, for Heidegger, such revelation in technology is fraught with threats, since technology, in the bringing-forth necessary for revelation, compels a particular type of human activity that is obscuring. This activity is the organizing and gathering that leads to seeing things, particularly nature, as that which is standing-reserve (*Bestand*). Whereas, in time past, material human creations that allowed one to interact with the world would lead to a constant recognition of the intimate relationship of one to the world, technology obscures this by making natural resources something to be stored for later use. This type of obscuring is particularly worrisome to Heidegger, since it goes beyond objectification of nature, to giving nature a perpetually non-object status while we wait to use it. This activity of gathering and organizing the world in terms of standing-reserve he terms,

---

[1] Heidegger, *Being and Time*, 196-211. For a helpful explanation of those aspects of truth that are relevant to technology, see Dilek Arh Çil, "The Relation between Technology and Truth in Heidegger's The Question concerning Technology," *Synthesis Philosophica* 53.1 (2012): 81-89.

"Enframing" (*Gestell*).[1] In Enframing, technological being is obscured, because the function of technology to reveal possibilities is reduced to being used for one possibility, which is the storing of resources or energy. In doing so, one's involvement with the world is downplayed, while technology gains the quality of being understood solely in terms of its instrumentality. Furthermore, this danger is universal in that technological being is fundamental to *Dasein,* and thus not just symptomatic of Western societies.[2]

To clarify how Heidegger understands technology, it is useful to see the way in which he analyzes actual technological objects. One of the examples he uses in "The Question Concerning Technology" is the hydroelectric dam on the Rhine, which he believes obscures one's relationship to the river.[3] In damming up a river for later use in producing electricity, one's possibilities in terms of that river is hidden under its water standing-reserve. Thus, the water is no longer the source of possibilities for the enactment of human being, but is something that is ignored until it is called on to produce. That the electricity produced by the dam is transmitted over long distances to urban areas, further exacerbates the relational discontinuity

---

[1] Heidegger, "Question concerning Technology," 19. The term *"Gestell"* is used, in German, for shelving. This likely points to the use of Enframing, which is necessary from Heidegger's perspective, as a constructed framework that aids enactment.

[2] For an example of argument for the universality of technological dangers, see Søren Riis, "Towards the Origin of Modern Technology: Reconfiguring Martin Heidegger's Thinking," *Continental Philosophy Review* 44.1 (2011): 103-117.

[3] Riis, "Origin of Modern Technology," 116.

between humans and the earth. Hence, the bringing-forth of technology is lost in the waiting to bring-forth.

This analysis of technology can be contrasted to Heidegger's description of a bridge over a river in his work "Building Dwelling Thinking."[1] For him, the bridge is significant in that it begins to define the relationship between various things that are in-the-world. Until one, in the course of one's life, needs to get to the other side of a river and thus builds a bridge, there are not, for her, banks to that river. Instead, there is simply a river that one may use, enjoy, etc. When, on the other hand, she desires to cross the river, she needs to span it with a bridge. In doing so, the banks become defined in terms of the structure that moves from one bank to the other. Previously, the bank had little or no conscious recognition, in that one's primal enactment of being would not likely require one to interpret the edge of the water as a bank. However, if a bridge is built, it forces a recognition of the edges of the water and thus defines the banks. In other words, for Heidegger, the reality of the banks inheres in their linguistic designation of relationship to the river, a relationship that is defined when one builds a bridge.[2] Until there is a bridge, there is only a river. Once there is a bridge, there is a way (i.e., a mode of enactment) to get from one bank to the other bank. Thus, the bridge clarifies, in Heidegger's mind, the relationship between humans, the earth, the air, and the gods (the last of these being a discussion for another

---

[1] Martin Heidegger, "Building Dwelling Thinking," *Basic Writings* (trans. Albert Hofstadter; ed. David Farrell Krell; New York: Harper Collins, 1993), 343-364.

[2] For Heidegger's most extended treatment of language, see Martin Heidegger, *Unterwegs zur Sprache* (Pfullingen: Neske, 1959).

time).[1] This is not to say that there was no river and banks until they were linguistically designated, but they had no reality for *Dasein* until they were spoken. Furthermore, in this naming, the river's essence is unveiled without being fundamentally altered. In the case of the dam, in contradistinction, the Rhine is now understood in terms of how it is used. Thus, Heidegger writes, "What the river is now, namely, a water power supplier, derives from out of the essence of the power station."[2] Here, technology has the potential to obscure the nature of things in the world by defining them through the technology, rather than allowing the technology (i.e. tool) to reveal their fundamental ontology. Simply put, whereas technology, such as a dam, often problematizes one's direct relationship to the world, tools, such as bridges, tend to clarify them.

Furthermore, the permanence and continuity of enactment that is designated by tools, is significant to the manner in which Heidegger's constructivist understanding of language comes together with his understanding of technology. The bridge's enduring usage gives rise to the persistent recognition of the river, its banks, the expanse it is spanning, the city to which one is headed, etc. The dam, on the other hand, encourages the forgetting of it and the Rhine in that Enframing places *Dasein* in the mode of ignoring them until they are needed. Heidegger writes, "But, it will be replied, the Rhine is still a river in the landscape, is it not? Perhaps, But how? In no other way than as an object on call for inspection by a tour group ordered there by

---

[1] See Heidegger, "Building Dwelling Thinking," 355ff.

[2] Heidegger, "Question concerning Technology," 16.

the vacation industry."[1] While the bridge constantly defines and provides the basis for language whereby the truth of relationality can be seen, the dam has the potential to do the contrary. Thus, language, which constructs reality,[2] gives voice to those relationships disclosed in *Dasein's* enactment. This constructivism gives rise to Heidegger's propensity to use *Wirklichkeit* terminology to speak of reality as opposed to *Realität*,[3] and the significance of the type of "dwelling" thinking that creates persisting artifacts (e.g. bridges) that help define *Dasein's* relationship to the world.

Heidegger's above quote concerning the Rhine's landscape is illuminating not only because of the atypical frustration that its author expresses, but also because it hints at his discussion of art, which bears similar

---

[1] Heidegger, "Question concerning Technology", 16.

[2] How Heidegger's view of reality relates to realism is difficult to discern. It has not been uncommon in German philosophy and theology to see the ideas as separable. Yet, realism tends to imply independence in that it holds that things can exist independently of observers. Yet, that the nature of *Dasein* as being-in-the-world ("*in-der-Welt-Sein*"), would prevent such a separation. Thus, Heidegger would appear to be reluctant to subscribe to realism if it was too associated with the objectifying talk that he feels only makes sense at an ontic level.

[3] Though both words are translated in the English with "reality", there are definite nuances. *Realität* usually has a more static backdrop, which is why the plural of the term is often translated "facts". Thus, it has closer affinities for terms like "realism" where the independent ontological status is recognized. *Wirklichkeit*, however, in being derived from the *wirken* semantic group, has a greater focus on enactment. Thus, it has a stronger relational connotation, in that *Wirklichkeit* is the way one enacts one's relationship to the world. In this way, it is more frequently used in constructivist contexts, whereby one's relational ontology with respect to the world is in view.

concerns. In "The Origin of the Work of Art,"[1] he develops an aesthetic that revolves around the manner in which a world is opened up to an observer whereby possibilities are expressed.[2] He comments on Van Gogh's A Pair of Shoes, wherein a pair of muddy work boots, which would otherwise be ignored, are brought to the attention of an onlooker such that she begins to question the fundamental relationship that the shoes possess in relationship to the worker. In this way, the tool's essence in enactment is retained even as the manner in which the tool is related to the user is contemplated. The art, then, gives permanency to the rupture structure necessary to force the type of ontological thinking necessary for the revelation of Being to occur. In that the work boots are no longer on a porch and ignored, but enshrined in a museum, their world is opened up as one's attention is disrupted and the relationship of the work boots to *Dasein* has the possibility of being explored.[3] Yet, this rupture structure does not, at least by Heidegger's account, occur at the expense of giving the shoes

---

[1] Martin Heidegger, "The Origin of the Work of Art," *Basic Writings* (trans. Albert Hofstadter; ed. David Farrell Krell, New York: Harper Collins, 1993), 139-212.

[2] In this, Heidegger's view has certain affinities with world projection theories. An example would be that developed in Nicholas Wolterstorff, *Works and Worlds of Art* (New York: Clarendon Press, 1980).

[3] It may be objected that the contemplation of *Dasein* as being-in-the-world is absent from most observation of art. However, even if *Dasein* is not explicit, the human existential condition represented in the artistic endeavor is frequently present. Hence, humans as representative of *Dasein* would not be necessarily absent. Also of significance is that one does not need to be in wholesale agreement with Heidegger's aesthetic views to admit of the world opening possibilities of art.

objective presence, as can be the case, he notices, when a hammer breaks[1] and is interpreted as an object that cannot be used. In the latter case, which will be expounded upon subsequently, the relationship of the hammer to Dasein's being is obscured. As is rarely recognized, the art not only provides the disruption in attention necessary for disclosure, but it has a persistence that provides continual reminder of one's need to understand things in their fundamental ontology. Whereas the dam on the Rhine invites one to ignore it until it is useful, Van Gogh's *Shoes* invites one to pay attention to it until truth is revealed.[2]

Art, like the bridge of "Building Dwelling Thinking," not only has permanence, but the type of permanence that encourages a constant recognition of those fundamental modes of enactment that are essential to *Dasein's* Being. Thus, it has certain parallels with another example that Heidegger gives concerning

---

[1] See Heidegger, *Being and Time*, 62-83.

[2] An intriguing implication of Heidegger's aesthetic perspective is that it would appear to make content significant in defining art *qua* art. The ability of the world of that which is depicted to be opened up in its fundamental ontology, intrinsically poses primal modes of enactment as the necessary subject matter of artistic endeavors. For example, a painting that contained fantasy elements may not provide the disruption necessary to force recognition of *Dasein*'s possibilities in *this* world, which would appear to preclude it from being art. What is also interesting is that the same might not be the case for the artistic representation of technology. Were the dam on the Rhine be painted and hung in a gallery as high art, it may serve the function of opening up possibilities to the observer. If this were the case, then one might be able to argue that art could serve to force certain aspects of tool use upon the dam in that it would, at the very least, provide the rupture structure necessary to prevent one from ignoring the Rhine, its dam, etc. This would seem to prevent, or at least mitigate, the effects of Enframing.

technology that can be illuminating rather than obscuring, which is a windmill. In that the source of the power of the windmill is not held in reserve, the constant relationship between the one employing the windmill and the world (i.e., wind) that powers it, allows for *Dasein's* Being to be disclosed.

Heidegger makes a similar claim earlier in his career with his analysis of tool use in *Being and Time*. When a hammer is understood as a hammer, it loses its ability to disclose something of Being, specifically the human activity of hammering. A hammer is only understood as a hammer when it is separated from the activity and becomes an independent object of thought. However, when a tool is being used, one does not interpret it, but simply enacts their being with it. Thus, a hammer can clarify being in showing the possibilities of human enactment, but it can obscure them also if the hammer is separated from the user (i.e. given objective presence). To summarize, technology, as something that is part of human activity and creativity, can give a glimpse into something significant of human being. It can disclose the possibilities of enactment that are previously concealed. Yet, the nature of technology is such that in the act of gathering to use for the purpose of disclosure, the act of gathering is emphasized until technology is reduced to its instrumentality in allowing individuals to organize the world into that which is standing-reserve. Consequently Heidegger writes:

> Both are ways of revealing, *alētheia.* In Enframing, that unconcealment comes to pass in conformity with which the work of modern technology reveals the real as standing-reserve. This work is therefore neither only a human activity nor a mere means within such activity.

> The merely instrumental, merely anthropological definition of technology is therefore in principle untenable.[1]

Therefore, technology which can disclose human possibilities can also conceal them. This occurs in its destining of humans towards gathering and organizing which leads to a view of technology that both emphasizes instrumental usage and also focuses on the world as something to hold in reserve until it is used.

**Areas of Danger**

An extended treatment of the use of Heidegger's criticism of technology would include a critique of his perspective[2] and the mapping of how it could be used across different belief systems. Yet, with limited space, this chapter will assume he is at least correct enough to be applicable to many forms of technology and will narrow the application to a Christian perspective (broadly construed). I will address four areas of danger.

The first problem posed by technology, were Heidegger's criticisms to be substantially correct, is the potential of the technological life obscuring one's relationship to the world and one's possibilities. By causing a disjunction between the user and what is being

---

[1] Heidegger, "Question concerning Technology," 3-5.

[2] A few such criticisms would include the need of an ethical substructure for speaking about technology in moral terms, an assumption of controversial views on ontology and metaphysics, the use of a constructivist view of language, and a close association with anti-realism. Furthermore, some may question his ability to analyze technology without giving a clear and concise definition of the term. It should be noted both that particular faith systems may not be at odds with much of Heidegger's perspective and that Christianity may actually be able to help remedy some of the difficulties in his overall viewpoint.

used, manifold negative outcomes occur. One that is of special significance, is that an individual's possibilities in relationship to particular enactments of human life may be made ambiguous. For example, a farmer may begin to see his land as something to be used rather than part of the world to which he is constantly related. That the life of farming, which is mentioned by Heidegger, requires a constant activity whereby one is reliant upon the land, points to it being a fundamental mode of enactment of *Dasein*. One might be able to argue, then, that farming is not the act of tilling the land for plant products, but a type of being that is possessed by humans that constantly seek to work within a dependency. Regardless of interpretation, the farmer, by the use of technology, could lose specificity in his relationship to the fields. Were, for example, robotic machinery able to do all the activities required for a crop to be yielded, then the land is no longer tended to constantly by the farmer as one who constantly depends on it, but is just a place to hold crops in reserve until harvest. Furthermore, it becomes a place that is ignored by the farmer, since only the robots have a constant connection to it.

This obscuring of one's relationship to the world has effects on Christianity. First, one would lose the force of agricultural metaphors in scriptural contexts, since the enactment of the farmer would be radically redefined by the instrument-oriented perspective of nature and technology. It would mean that to be a "vine-dresser,"[1] or for a tree to "bear fruit,"[2] or for a congregation to not "muzzle the ox,"[3] would not have the proper

[1] John 15:1.

[2] Matthew 7:17ff.

[3] 1 Timothy 5:18.

interpretive horizon as one's relationship to the world is unclear. In Heideggerean terminology, the language of fundamental ontology may be lost, so much so that one would have obstacles in reaching past the ontic activities to the question of being. In Christian terms, the loss of one's relationship to the world could prevent the metaphors of the Bible from having their cognitive and affective power in opening up believers to their possibilities. Second, the misunderstanding that the world is not something to which one is intimately related could map on one's understanding of what it means to "abide"[1] in Christ. That the world is an object to be studied or a resource to be stored and forgotten, means that it is not something to which one must be related. Consequently, if knowledge of Christ is mediated, then obscuring the relationship with the world can conceal those modes of Christ being conveyed to individuals in the world. This means that a world that is separable from *Dasein* may prevent an understanding of Christ where he is inseparable from the believer.

Second, Heidegger's view of technology may simplify one's relationship to the world in that the usage necessary for the technological life may lead to the understanding of the world in its instrumentality. Thus, the world, could be understood in terms of how it could be used and not in terms of how it relates to the totality of human enactment. This could lead to the further difficulty of seeing other humans in their instrumentality, which would have drastic ethical implications. The historical Christian perspective on love prevents seeing others, including God, in terms of how one can use them for one's own ends. It is likely this

[1] John 15:4-16.

perspective that Kant appropriates from his Lutheran pietism, when he attempts to prevent humans from being treated as means. Treating others as instruments is also reinforced by the detachment of relationships that the technological life can produce, since one would fail to see themselves as having an intimate relationship with the individual to be used. Others, in this way, could be seen as standing reserve, an idea to which Heidegger alludes in his explanation of "human resources."[1] In essence, Christians must be concerned with the ethical implications of people increasingly seeing others as something to be used with technology preventing the user from having to look them in the eyes when using them.

Third, beyond the previous concerns, technology, from a Heideggerean perspective, can make the prospects of revelation precarious. Where technology should encourage a phenomenological mode so that one asks, "What is being revealed here?" one instead begins to ask, "How can this be used?" Thus, those things that are meant to reveal become that which should be employed. Not only can this fundamentally close off the type of content which should be observed and analyzed, but it could also have the severe consequence of distorting the revelatory activity that is foundational to the Christian faith. The mode of receptivity inherent in the idea of revelation may be modified such that it would not be able to perform the conceptual function within theology in relating humans to their dependence on God. In essence, technology can place emphasis on actively using the world and not passively seeing truth and revelation in it. In encouraging this mode of being, it

[1] Heidegger, "Question Concerning Technology," 18.

also may problematize the relationship of the Christian to whatever is considered to be the revelation of God. Were the nature of that activity (i.e. revelation) to be lost, then what one sees Christ or Scripture to be doing could be significantly altered.

Lastly, technology may complicate the ability of language to be the foundation of Christian thought and proclamation. Language, which is the way being can be revealed in voicing of the possibilities of the human life, can be altered as those possibilities are themselves changed. Similar to the Wittgensteinian observation that a change in a form of life can have linguistic effects, the technological life that obscures possibilities may change language to such a degree that the proclamation of the Word of God may become increasingly difficult. If Heidegger is right, then a dam could prevent a river from being a river to us. Instead, a river becomes a source of power, a resource, etc. In this way, the essence of the river can be obscured and the term "river" could be redefined. That the dam can alter one's understanding of the river means that technology can modify the discourse that describes *Dasein's* relationship to the world. In this way, technology could change the way in which words reveal being, such that the Christian speech that is required for evangelism, catechism, discipleship, etc., would be altered. For some, this may not be problematic, but if revelation from God is an intentional activity, and this activity is interpreted in words, the alteration of the words could make it difficult to recognize divine intention.

**Conclusion**

The relationship of the Christian faith to technology is one in which the ends of the spectrum of viewpoints are radically different. On one hand, you see Christian types of transhumanism that speak of technology in messianic terms. On the other, you have individuals of the Christian faith that resist most forms of technology, except in the cases where it can be equated with tools that have Scriptural representation. Heidegger's view is more moderate in that he sees both potentialities and problems within technological living. His view of technology as indicative of being which can reveal human possibilities comes with the incumbent danger that technological being simplifies human enactment into gathering into a standing-reserve. If Heidegger is substantially correct, then this would lead to a strongly instrumental view of technology that has perils in terms of understanding relationships, of an instrumental approach to interpersonal relationships, and of the altering of discourse necessary for the Christian life and mission.

## *Contributors*

***John Becker*** is a Part-time Lecturer of Theological Studies at Loyola Marymount University in Los Angeles, CA. He earned his PhD in Comparative Theology and Philosophy from Claremont School of Theology and specializes in Buddhist-Christian studies in dialogue with process thought.

***Dean G. Blevins*** is Professor of Practical Theology and Christian Discipleship and Director of the Masters of Arts in Christian Formation and Discipleship at Nazarene Theological Seminary. Prior to NTS, Dr. Blevins served eleven years at Trevecca Nazarene University as Professor of Christian Education and also served as the inaugural J.B. Elizer Chair of Christian Ministry.

***Greg Cootsona*** is Lecturer. Religious Studies and Humanities. California State University at Chico, 2014. Besides his interest in religion and science, he has scholarly activities centering on C.S. Lewis.

***John Culp*** is Professor of Philosophy Emeritus at Azusa Pacific University. John holds a BA in Religion and Philosophy from Greenville College, MDiv from Asbury Theological Seminary, MA in Theology from Butler University, and PhD in Philosophy of Religion/ Theology from Claremont Graduate School.

***Rem B. Edwards*** is Lindsay Young Professor of Philosophy, Emeritus, at the University of Tennessee, Knoxville. He has published 21 books and over 100 articles and reviews. His most recent books are: *John Wesley's Values – And Ours* (2012), *Spiritual Values and Evaluations* (2012), and *An Axiological Process Ethics* (2014). In graduate school, he was a Danforth Graduate Fellow and received a BD degree from Yale Divinity School and a PhD in Philosophy from Emory University.

***Matthew Nelson Hill*** is Associate Professor of Philosophy and E.P. Hart Honors Director at Spring Arbor University. His recent books are *Evolution and Holiness* (IVP-Academic, 2016) and *In Spirit and In Truth* (Claremont School of Theology Press, 2016). His research interests include faith and science, evolution, and philosophy.

***Wm. Curtis Holtzen*** is Professor of Philosophy and Theology at Hope International University. His other books include *In Spirit and In Truth: Philosophical Reflections on Worship and Liturgy* (Claremont School of Theology Press, 2016) and *By Faith and Reason: The Essential Keith Ward* (Darton, Lomgman, & Todd, 2012). Research interests include the nature of faith, trust, and hope as well as the attributes of God.

***Joshua Kira*** a Kira is Assistant Professor of Philosophy and Theology at Cedarville University where he teaches theology courses both in Bible and philosophy. His current research interests include Heideggerian phenomenology, philosophy of language, theology of revelation, and theology of being.

*Joyce Ann Konigsburg* is Visiting Assistant Professor of Religious Studies at Notre Dame of Maryland University. She is also a Founding Fellow of the New York City think tank, Sinai and Synapses, which explores the relationship between religion and science, which is one of her research interests, along with interreligious studies and systematic theology.

*Mark H. Mann* is Associate Professor of Theology and director of the Honors Scholars Program at Point Loma Nazarene University in San Diego, CA.

*Isaac Wiegman* is a Lecturer of Philosophy at Texas State University who studies human nature and its normative significance, with much of this work focusing on anger, revenge, and retribution. Recent publications include "The Evolution of Retribution: Intuitions Undermined" and "Angry Rats and Scaredy Cats: Lessons from Competing Cognitive Homologies."

# *Bibliography*

Alexander, Eben. *Proof of Heaven: A Neurosurgeon's Journey into the Afterlife*. New York: Simon & Schuster, 2012.

Allais, Lucy. "Wiping the Slate Clean: The Heart of Forgiveness." *Philosophy and Public Affairs* 36.1 (2008): 33–68.

Aquinas, Thomas. *The Summa Contra Gentiles of Saint Thomas Aquinas*. New York: Benziger Bros., 1924.

_______. *The Summa Theologiae of St. Thomas Aquinas*. London: Burns Oates & Washbourne, 1912.

_______. *On Evil*. Translated by Richard Regan. Edited with an Introduction and Notes by Brian Davies. Oxford: Oxford University Press, 2003.

_______. *Summa Theologica*. Translated by Fathers of the English Dominican Province. Great Books series. Chicago: Encyclopedia Britannica, 1952.

Audi, Robert, ed. *The Cambridge Dictionary of Philosophy*. 2nd ed. Cambridge: Cambridge University Press, 1999.

Augros, Robert and George Stanciu. *The New Story of Science*. Washington DC: Regnery, 1984.

Augustine. *Concerning the City of God against the Pagans*. London: Penguin Books, 2003.

_______. *Confessions*. Translated by Richard Geoffrey Pine-Coffin. Harmondsworth: Penguin, 1961.

Barash, David P. and Judith Eve Lipton. *Payback: Why We Retaliate, Redirect Aggression, and Take Revenge*. Oxford: Oxford University Press, 2011.

Barber, Daniel Colucciello. *Deleuze and the Naming of God: Post-Secularism and the Future of Immanence*. Edinburgh: Edinburgh University Press, 2014.

Barbour, Ian. *Religion and Science: Historical and Contemporary Issues*. Revised Edition. San Francisco: HarperOne, 1997.

Baron, Jonathan and Ilana Ritov. "The Role of Probability of Detection in Judgments of Punishment." *SSRN Electronic Journal* 1.2 (2009): 553–90.

_______. "Intuitions about Penalties and Compensation in the Context of Tort Law." *Journal of Risk and Uncertainty* 7.1 (1993): 17–33.

Barrett, Justin L. *Born Believers: The Science of Children's Religious Belief*. New York: Free Press, 2012.

_______. *Cognitive Science, Religion and Theology: From Human Minds to Divine Minds*. Templeton Science and Religion Series. West Conshohocken: Templeton Press, 2011.

Barth, Karl. *A Shorter Commentary on Romans*. Translated by David. H. van Daalen. Edited by T. F. Torrance. Richmond: John Knox, 1959.

_______. *Church Dogmatics*. Translated by Geoffrey W. Bromiley. Edinburgh: T & T Clark, 1957.

Beauregard, Mario. *Brain Wars: The Scientific Battle over the Existence of the Mind and the Proof That Will Change the Way We Live Our Lives*. New York: HarperOne, 2012.

Beauregard, Mario, and Denyse O'Leary. *The Spiritual Brain: A Neuroscientist's Case for the Existence of the Soul*. New York: HarperOne, 2007.

Behe, Michael. *Darwin's Black Box*. New York: Free Press, 1996.

_______. *The Edge of Evolution*. New York: Free Press, 2007.

Beilby, James K. *Naturalism Defeated? Essays on Plantinga's Evolutionary Argument against Naturalism.* Ithaca: Cornell University Press, 2002.

Benson, Herbert, Jeffery A. Dusek, Jane B. Sherwood, Peter Lam, Charles F. Bethea, William Carpenter, Sidney Levitsky, Peter C. Hill, Donald W. Clem Jr., Manoj K. Jain, David Drumel, Stephen L. Kopecky, Paul S. Mueller, Dean Marek, Sue Rollins, and Patricia L. Hibberd. "Study of the Therapeutic Effects of Intercessory Prayer (STEP) in Cardiac Bypass Patients: A Multicenter Randomized Trial of Uncertainty and Certainty of Receiving Intercessory Prayer." *American Heart Journal* 151.4 (2006): 934-42.

Berger, Peter, and Thomas Luckman. *The Social Construction of Reality: A Treatise in the Sociology of Knowledge.* New York: Anchor Books, 1966.

Berns, Walter. *For Capital Punishment: Crime and the Morality of the Death Penalty.* New York: Basic Books, 1979.

Berry, Robert. J. "John Ray: Father of Natural Historians." *Science and Christian Belief* 13.1 (2001): 25-38.

Biernacki, Loriliai, and Philip Clayton, eds. *Panentheism across the World's Traditions.* Oxford: Oxford University Press, 2014.

Blakemore, Sarah-Jane and Uta Firth. *The Learning Brain: Lessons for Education.* Malden: Blackwell Publishing, 2005.

Blevins, Dean G. "Psychology of Religion." *Encyclopedia of Christian Education,* 1007-08. Edited by George Thomas Kurian and Mark A. Lamport. Lanham: Rowman and Littlefield, 2015.

Boundry, Maarten, Stefaan Blancke, and Johan Braeckman. "How Not to Attack Intelligent Design Creationism: Philosophical Misconceptions About Methodological Naturalism." *Foundations of Science* 15.3 (2010): 227–44.

Bracken, Joseph A., "Panentheism and the Classical God-World Relationship: A Systems-Oriented Approach." *American Journal of Theology & Philosophy* 36.3 (2015): 207-25.

_______. "Panentheism in the Context of the Theology and Science Dialogue." *Open Theology* 1 (2014): 1-11.

Brierley, Michael W. "The Potential of Panentheism for Dialogue between Science and Religion." *The Oxford Handbook of Religion and Science*, 635-51. Edited by Philip Clayton and Zachary Simpson. Oxford: Oxford University Press, 2006.

Brown, Brene. *Rising Strong: The Reckoning. The Rumble. The Revolution.* New York: Spiegel, 2015.

Brown, Warren S. "Resonance: A Model for Relating Science, Psychology, and Faith." *Journal of Psychology and Christianity* 23.2 (2004): 110-20.

_______. "The Brain, Religion, and Baseball: Comments on the Potential of a Neurology of Religion and Religious Experience." *Where God and Science Meet.* 3 vols. 2: 229-44. Edited by Patrick McNamara. Westport: Praeger Publishers, 2006.

Browning, Don. *Religious Thought and the Modern Psychologies.* 2nd ed. Minneapolis: Augsburg Fortress, 2000.

Burpo, Todd. *Heaven is for Real: A Little Boy's Astounding Story of His Trip to Heaven and Back*. Nashville: Thomas Nelson, 2011.

Burrell, David. *Faith and Freedom: An Interfaith Perspective*. Malden: Blackwell Publishing, 2004.

Bushman, Brad J., Angelica M. Bonacci, William C. Pedersen, Eduardo A. Vasquez, and Norman Miller. "Chewing on It Can Chew You Up: Effects of Rumination on Triggered Displaced Aggression." *Journal of Personality and Social Psychology* 88.6 (2005): 969–83.

Calvin, John. *Institutes of the Christian Religion*. Library of Christian Classics 20. 2 vols. Edited by John T. McNeil. Translated by Ford Lewis Battles. Philadelphia: Westminster, 1960.

Capps, Walter H. *Religious Studies: The Making of a Discipline*. Minneapolis: Fortress Press, 2000.

Carlsmith, Kevin M. "On Justifying Punishment: The Discrepancy Between Words and Actions." *Social Justice Research* 21.2 (May 2008): 119–37.

_______. "The Roles of Retribution and Utility in Determining Punishment." *Journal of Experimental Social Psychology* 42.4 (July 2006): 437–51.

Carlsmith, Kevin M. and John M. Darley. "Psychological Aspects of Retributive Justice." *Advances in Experimental Social Psychology* 40 (2008): 194-236.

Carlsmith, Kevin M., John M. Darley, and Paul H. Robinson. "Why Do We Punish?: Deterrence and Just Deserts as Motives for Punishment." *Journal of Personality and Social Psychology* 83.2 (2002): 284–

Chandrasekhar, Subrahmanyan. *Truth and Beauty: Aesthetic and Motivations in Science.* Chicago: University of Chicago Press, 1987.

Çil, Dilek Arh. "The Relation between Technology and Truth in Heidegger's The Question concerning Technology." *Synthesis Philosophica* 53.1 (2012): 81-89.

Clayton, Philip. "Conceptual Foundations of Emergence Theory." *The Re-Emergence of Emergence: The Emergentist Hypothesis from Science to Religion*, 1-31. Edited by Philip Clayton and Paul Davies. Oxford: Oxford University Press, 2006.

_______. "God and World." *The Cambridge Companion to Postmodern Theology*, 203-218. Edited by Kevin J. Vanhoozer. Cambridge: Cambridge University Press, 2003.

_______. "Kenotic Trinitarian Panentheism." *Dialog* 44.3 (Fall 2005): 250-55.

________. "Panentheist Internalism: Living within the Presence of the Trinitarian God," *Dialog* 40.3 (Fall 2001): 210.

Clayton, Philip and Steven Knapp, *The Predicament of Belief: Science, Philosophy, and Faith.* Oxford: Oxford University Press, 2013.

Clutton-Brock, Timothy H., and Geoff A. Parker. "Punishment in Animal Societies." *Nature* 373.19 (1995): 209–16.

Cooper, John W. "Panentheism: The Other 'God of the Philosophers': An Overview." *American Theological Inquiry* 1.1 (2008): 11-24.

_______. *Panentheism The Other God of the Philosophers: From Plato to the Present*. Grand Rapids: Baker Academic, 2006.

Cootsona, Gregory. *C. S. Lewis and the Crisis of a Christian*. Louisville: Westminster John Knox, 2014.

Cranfield, Charles E. B. *The Epistle to the Romans*. 2 vols. International Critical Commentary. Edinburgh: T & T Clark, 1975.

Crisp, Oliver D. "Divine Retribution: A Defence." *Sophia* 42.2 (October 2003): 35–52.

Culp, John. "Panentheism." *The Stanford Encyclopedia of Philosophy* (Summer 2015 Edition). Edited by Edward N. Zalta. http://plato.stanford.edu/archives/sum2015/entries/panentheism/.

D'Arms, Justin, and Daniel Jacobson. "Sentimentalism and Scientism." *Moral Psychology and Human Agency: Philosophical Essays on the Science of Ethics*, 253–78. Edited by Justin D'Arms and Daniel Jacobson. Oxford: Oxford University Press, 2014.

Daly, Martin, and Margo Wilson. *Homicide*. London: Transaction Publishers: 1988.

Damasio, Antonio. *Self Comes to Mind: Constructing the Conscious Brain*. New York: Pantheon Books, 2010.

Daniels, T. Scott. "Passing the Peace: Worship That Shapes Nonsubstitutionary Convictions." *Atonement and Violence: A Theological Conversation*, 125–48. Edited by John Sanders, Nashville: Abingdon, 2005.

Darley, John M., Kevin M. Carlsmith, and Paul H. Robinson. "Incapacitation and Just Deserts as Motives for Punishment." *Law and Human Behavior* 24.6 (2000): 659–83.

Dawkins, Richard. *The God Delusion*. New York: Mariner. 2008.

DeScioli, Peter, and Robert Kurzban. "Mysteries of Morality." *Cognition* 112.2 (August 2009): 281–99.

Dittrich, Luke. "The Prophet." *Esquire* 150.4 (December 2012): 88-128.

Douglas, Mary. *Leviticus as Literature*. Oxford: Oxford University Press, 1999.

Duff, Anthony. *Punishment, Communication, and Community*. Oxford: Oxford University Press, 2001.

Duhem, Pierre. *The Aim and Structure of Physical Theory*. Translated by Philip P. Wiener. Princeton: Princeton University Press, 1954.

Dukas, Helen and Banesh Hoffmann. *Albert Einstein: Creator and Rebel*. New York: Penguin, 1972.

Dunn, James D. G. *Romans 1-8*. Word Biblical Commentary 38A. Dallas: Word, 1988.

Dyck, Drew. "Study the Brain Without Losing Your Soul." *Leadership Journal* 35.3 (Summer 2014): 5.

Edwards, Rem Blanchard. "The Human Self: An Actual Entity or a Society?" *Process Studies* 5 (1975): 195-203.

_______. "Whitehead's Theistic Metaphysics and Axiology." *Process Studies* 45:1 (2016): 5-32.

_______. *What Caused the Big Bang?* Amsterdam: Rodopi, 2001.

Ellis, George. *Faith, Hope, and Doubt in Times of Uncertainty*. Queensland, Australia: Interactive Publications, 2008.

Enns, Peter. *Inspiration and Incarnation: Evangelicals and the Problem of the Old Testament*. Grand Rapids: Baker, 2005.

_______. *The Sin of Certainty: Why God Desires Our Trust More than Our "Correct" Belief*. New York: HarperOne, 2016.

Faber, Roland. "Der transreligiöse Diskurs. Zu einer Theologie transformativer Prozesse." *Polylog* 9 (2002): 65-94.

Farah, Martha. J., ed. *Neuroethics: An Introduction with Readings*. Cambridge: MIT Press, 2010.

Fehr, Ernst, and Simon Gächter. "Altruistic Punishment in Humans." *Nature* 415.6868 (January 2002): 137–40.

Feyerabend, Paul. *Against Method: Outline of an Anarchist Theory of Knowledge.* London: New Left Books, 1975.

_______. *Science in a Free Society.* London: New Left Books, 1978.

Fitzmyer, Joseph. *Romans: A New Translation and Commentary.* Anchor Bible. New York: Doubleday, 1992.

Forrest, Barbara. "Methodological Naturalism and Philosophical Naturalism: Clarifying the Connection." *Philo* 3.2 (2000): 7-29.

Frank, Robert H. *Passions Within Reason: The Strategic Role of the Emotions.* New York: Norton, 1988.

Frankenberry, Nancy. "Enduring Questions in Philosophy of Religion: A Response to Neville and Godlove." *American Journal of Theology & Philosophy* 37.1 (2016): 36-52.

Freud, Sigmund. *Civilization and its Discontents.* Standard Edition. New York: W.W. Norton, 1989 (1930).

Gadamer, Hans-Georg. *Truth and Method.* 2nd Edition. Translated by Joel Weinsheimer and Donald G. Marshall. New York: Continuum, 1989.

García-Rivera, Alejandro. *The Community of the Beautiful: A Theological Aesthetics.* Collegeville: Liturgical Press, 1999.

Gay, Volney P. "Introduction." *Neuroscience and Religion: Brain, Minds, Self, and Soul,* 1-15. Edited by Volney P. Gay. Lanham: Lexington Books, 2009.

Gocke, Benedikt Paul. "Panentheism and Classical Theism." *Sophia* 52.1 (April 2013): 61-75.

Green, Clifford, ed. *Karl Barth: Theologian of Freedom.* London: Collins: 1989.

Green, Joel B. *Body, Soul, and Human Life: The Nature of Humanity in the Bible*. Grand Rapids: Baker Academic, 2008.

_______. ed. *In Search of the Soul: Four Views of the Mind-Body Problem*. Eugene: Wipf and Stock, 2005.

Gregersen, Niels Henrik. "Three Varieties of Panentheism." *In Whom We Live and Move and Have Our Being: Panentheistic Reflections on God's Presence in a Scientific* World, 19-35. Edited by Philip Clayton and Arthur Peacocke. Grand Rapids: Eerdmans, 2004.

Griffin, David Ray. "Panentheism: A Postmodern Revelation." *In Whom We Live and Move and Have Our Being*, 36-47. Edited by Philip Clayton and Arthur Peacocke. Grand Rapids: Eerdmans, 2004.

_______. *Panentheism and Scientific Naturalism: Rethinking Evil, Morality, Religious Experience, Religious Pluralism, and the Academic Study of Religion*. Claremont, CA: Process Century Press, 2014.

_______. "Process Eschatology." *The Oxford Handbook of Eschatology*, 295-310. Edited by Jerry L. Walls. Oxford: Oxford University Press, 2008.

_______. "Process Philosophy of Religion." *International Journal for Philosophy of Religion* 50 (2001): 131-151.

_______. *Religion and Scientific Naturalism: Overcoming the Conflicts*. Albany: State University of New York Press, 2000.

Griffiths, Paul E. "Basic Emotions, Complex Emotions, Machiavellian Emotions." *Philosophy and the Emotions*, 39–67. Edited by Anthony Hatzimoysis. Cambridge: Cambridge University Press, 2003.

Hagerty, Barbara Bradley. *Fingerprints of God: The Search for the Science of Spirituality*. New York: Riverhead/Penguin Books, 2009.

Hall, Douglas. John. *The Steward: A Biblical Symbol Come of Age*. Grand Rapids: Eerdmans, 1990.

_______. *The Steward: A Biblical Symbol Come of Age*. Rev. ed. Eugene: Wipf & Stock, 2004.

Hardiman, Mariale. M. *Connecting Brain Research with Effective Teaching: The Brain-Targeted Teaching Model*. Lanham: Scarecrow Education, 2003.

Hartshorne, Charles. *Omnipotence and other Theological Mistakes*. Albany: State University of New York Press, 1984.

_______. *The Divine Relativity: A Social Conception of God*. New Haven: Yale University Press, 1948.

Hartshorne, Charles and William L. Reese. *Philosophers Speak of God*. Chicago: University of Chicago

Heidegger, Martin *Being and Time*. Translated by Joan Stambaugh. New York: State University of New York Press, 1996.

________. "Building Dwelling Thinking." *Basic Writings*, 343-364. Translated by Albert Hofstadter. Edited by David Farrell Krell. New York: Harper Collins, 1993.

________. *Sein und Zeit*. Tübingen: Max Niemeyer Verlag, 1956.

________. "The Origin of the Work of Art." *Basic Writings*, 139-212. Translated by Albert Hofstadter. Edited by David Farrell Krell. New York: Harper Collins, 1993.

________. "The Question concerning Technology." *The Question concerning Technology and Other Essays*, 3-35. Translated by William Lovitt. New York: Garland Publishing, Inc., 1977.

________. *Unterwegs zur Sprache*. Pfullingen: Neske, 1959.

Hopkins, Gerard Manley. "The Golden Echo." *Selected Poems of Gerard Manley Hopkins*, 48-49. Edited by Bob Blaisdell. Mineola: Dover, 2011.

Hurka, Thomas. "The Common Structure of Virtue and Desert." *Ethics* 112.1 (2001): 6–31.

Husserl, Edmund. *The Crisis of European Sciences and Transcendental Phenomenology*. Translated by David Carr. Evanston: Northwestern University Press, 1970.

Jantzen, Grace. *God's World, God's Body*. Philadelphia: Westminster Press, 1984.

Jeeves, Malcom and Warren. S. Brown. *Neuroscience, Psychology, and Religion: Illusions, Delusions, and Realities about Human Nature.* West Conshohocken: Templeton Foundation Press, 2009.

Jeffreys, Derek S. "How Reformed Is Reformed Epistemology? Alvin Plantinga and Calvin's '*Sensus Divinitatis*.'" *Religious Studies* 33.4 (1997): 419-31.

Jensen, Eric. *Brain-based Learning: The New Paradigm of Teaching.* Thousand Oaks: Corwin Press, 2008.

Johnson, Elizabeth A. "Does God Play Dice? Divine Providence and Chance." *Theological Studies* 56 (1996): 3–18.

Jones, Charles B. *Introduction to the Study of Religion*. The Great Courses Audio Series. Chantilly VA: The Teaching Company, 2007.

________. *The View from Mars Hill: Christianity in the Landscape of World Religions*. Cambridge: Cowley Press, 2005.

Kaipayil, Joseph. *Relationalism: A Theory of Being*. Bangalore: Jeevalaya Institute of Philosophy Publications, 2009.

Kearney, Richard. *Anatheism: Returning to God after God*. New York: Columbia University Press, 2011.

Keltner, Dacher. "Signs of Appeasement: Evidence for the Distinct Displays of Embarrassment, Amusement, and Shame." *Journal of Personality and Social Psychology* 68.3 (1995): 441–54.

Komasinski, Andrew. "Anti-Climacus's Pre-emptive Critique of Heidegger's 'Question Concerning Technology.'" *International Philosophical Quarterly* 54.3 (2014): 265-77.

Kuhn, Thomas. *The Structure of Scientific Revolutions.* Chicago: University of Chicago Press, 1962.

Lakatos, Imre. "Falsification and the Methodology of Scientific Research Programmes." *Criticism and the Growth of Knowledge*, 91-106. Edited by Imre Lakatos and Alan Musgrave. Cambridge: Cambridge University Press, 1970.

Laudan, Larry. "The Demise of the Demarcation Problem." *Physics, Philosophy and Psychoanalysis: Essays in Honor of Adolf Grunbaum*, 111-27. Edited by Robert Sonné Cohen and Larry Laudan. Boston: Reidel, 1983.

Lewis, C. S. *God in the Dock: Essays on Theology and Ethics*. Edited by Walter Hooper. Grand Rapids: Eerdmans, 1970.

_______. *Mere Christianity*. New York: MacMillan, 1960.

_______. *The Problem of Pain*. New York: MacMillan, 1962.

Lewis, David. "Do We Believe in Penal Substitution?" *Philosophical Papers* 26.3 (1997): 203-09.

Lloyd-Jones, Sally. *The Jesus Story Book Bible*. Grand Rapids: Zondervan, 2007.

Lodahl, Michael. *God of Nature and of Grace: Reading the World in a Wesleyan Way*. Nashville: Kingswood, 2004.

Loder, James. E. *The Logic of the Spirit: Human Development in Theological Perspective*. San Francisco: Jossey-Bass 1998.

_______. *The Transforming Moment.* 2nd Edition. Colorado Springs: Helmers and Howard, 1989.

Loder, James. E. and W. Jim Neidhardt. *The Knight's Move: The Relational Logic of the Spirit in Theology and Science.* Colorado Springs: Helmers and Howard, 1992.

Lonergan, Bernard J. F. *Insight: A Study of Human Understanding.* Edited by Frederick E. Crowe and Robert M. Doran. Toronto: University of Toronto, 1992.

Lossky, Vladimir. *The Mystical Theology of the Eastern Church.* New York: St. Vladimir's Seminary, 1997.

Ma, Lin and Jaap van Brakel. "Heidegger's Thinking on the 'Same' of Science and Technology." *Continental Philosophy Review,* 47.1 (2014): 19-43.

Maddox, Randy. "John Wesley's Precedent for Theological Engagement with the Natural Sciences." *Wesleyan Theological Journal* 44.1 (Spring 2009): 23-54.

_______. "The Enriching Role of Experience." *Wesley and the Quadrilateral: Renewing the Conversation.* 107-128. Edited by W. Stephen Gunter. Nashville: Abingdon Press, 1997.

Malony, H. Newton, ed. *Psychology of Religion: Personalities, Problems, Possibilities.* Grand Rapids: Baker Book House, 1991.

Mann, Mark and Ron Benefiel. "Our Wesleyan Tradition: Wesleyan Faith and Practice and the PLNU Mission." *Didache: Faithful Teaching* 12:2 (Winter 2013): http://didache.nazarene.org/index.php/volume-12-2/878-didache-v12n2-01-our-wesleyan-tradition-plnu/file.

Matthews, Rex. "Religion and Reason Joined: A Study in the Theology of John Wesley." Harvard Th.D. Thesis, 1986.

McAllister, James W. "Truth and Beauty in Scientific Reason." *Synthese* 78.1 (1989): 25-35.

McFague, Sallie. *Models of God: Theology for an Ecological, Nuclear Age*. Philadelphia: Fortress Press, 1987.

McGrath, Alister. *A Scientific Theology, Volume 1: Nature*. Grand Rapids: Eerdmans, 2001.

________. *Science and Religion: A New Introduction*. Oxford: Wiley-Blackwell, 2010.

McMullin, Ernan. "Plantinga's Defense of Special Creation." *Christian Scholar's Review* 21.1 (1991): 55-79.

McNamara, Patrick. *The Neuroscience of Religious Experience*. New York: Cambridge University Press, 2009.

*Methodist Book of Discipline*. Nashville: United Methodist Publishing House, 2012.

Milbank, John, Catherine Pickstock and Graham Ward, eds. *Radical Orthodoxy: A New Theology*. London: Routledge, 1999.

________. *Theology and Social Theory: Beyond Secular Reason*. 2nd ed. New York: Wiley and Sons, 2006.

Miles, Rebekah L. "The Instrumental Role of Reason." *Wesley and the Quadrilateral: Renewing the Conversation*, 77-106. Edited by W. Stephen Gunter. Nashville: Abingdon Press, 1997.

Miller, Keith B. "Design and Purpose within an Evolving Creation." *Darwinism Defeated? The Johnson-Lamoureux Debate on Biological Origins*, 109-20. Edited by Phillip E. Johnson, Denis O. Lamoureux, *et al.* Vancouver: Regent College Publishers, 1999.

Mitchell, Louis J. *Jonathan Edwards and the Experience of Beauty*. Studies in Reformed Theology and History 9. Princeton: Princeton Theological Seminary Press, 2003.

Moltmann, Jurgen. *God in Creation: A New Theology of Creation and the Spirit of God*. Translated by Margaret Kohl. San Francisco: Harper and Row, 1985.

_______. *The Coming of God: Christian Eschatology*. Translated by Margaret Kohl. Minneapolis: Fortress, 1996.

Moore, Michael S. *Placing Blame: A Theory of the Criminal Law*. Oxford: Oxford University Press, 2010.

Murphy, Mark C. "Not Penal Substitution but Vicarious Punishment." *Faith and Philosophy* 26.3 (2009): 253–73.

Murphy, Nancey. *Bodies and Souls, or Spirited Bodies?* New York: Cambridge University Press, 2006.

_______. *Theology for a Scientific Age*. Ithaca: Cornell University Press, 1993.

Nash, Ronald H. "Process theology and Classical Theism." *Process Theology*, 3-29. Edited by Ronald H. Nash. Grand Rapids: Baker Book House, 1987.

_______. *The Concept of God*. Grand Rapids: Zondervan, 1983.

Nelissen, Rob M. A. and Marcel Zeelenberg. "Moral Emotions as Determinants of Third-Party Punishment: Anger, Guilt, and the Functions of Altruistic Sanctions." *Judgment and Decsion Making* 4.7 (2009): 543–53.

Newberg, Andrew B. *Principles of Neurotheology*. Ashgate Religion and Science Series. Burlington: Ashgate, 2010.

Newberg, Andrew and Eugene D'Aquili. *Why God Won't God Away: Brain Science and the Biology of Belief.* New York: Ballantine, 2002.

Newberg, Andrew, and Mark Robert Waldman. *Born to Believe: God, Science, and the Origin of Ordinary and Extraordinary Beliefs.* New York: Free Press, 2007.

Nikkel, David H. "Panentheism." *Encyclopedia of Science and Religion*, 641–45. Edited by J. Wentzel Vrede Van Huyssteen. New York: MacMillan Reference USA, 2003.

Nisbett, Richard E., and Dov Cohen. *Culture of Honor: The Psychology of Violence in the South.* Boulder: Westview Press, 1996.

Noro, Yoshio. "Wesley's Theological Epistemology." *The Iliff Review* 28 (1971): 59-67.

Nowak, Martin A., Karen M. Page and Karl Sigmund. "Fairness versus Reason in the Ultimatum Game." *Science* 289.5485 (September 2000): 1773–75.

O'Connor, Cailin. "Guilt, Games, and Evolution." *Emotion Researcher*. Edited by Andrea Scarantino. (2016): http://emotionresearcher.com/guilt-games-and-evolution/.

Oden, Thomas C. *The Living God.* San Francisco: Harper-SanFrancisco, 1987.

Oestricher, Mark. "Neuron-Shmeuron: Why Should We Care About Adolescent Brain Development?" *Youthworker Journal* 30.5 (May/June 2014): http://www.youthworker.com/neuron-hmeuron-why-should-we-care-about-adolescent-brain-development/ (accessed 4/11/17).

Olson, Roger E. "A Postconservative Evangelical Response to Panentheism." *Evangelical Review of Theology* 85.4 (2013): 328-37.

Oord, Thomas Jay. *The Uncontrolling Love of God: An Open and Relational Account of Providence*. Downers Grove: IVP Academic, 2015.

_______. *Defining Love*. Grand Rapids: Brazos Press, 2010.

_______. *The Nature of Love: A Theology*. St. Louis: Chalice Press, 2010.

Ormerod, Neil. "Bernard Lonergan and the Recovery of a Metaphysical Frame." *Theological Studies* 74 (2013): 960–82.

Orsi, Robert A. "Introduction." *The Cambridge Companion to Religious Studies*, 1-15. New York: Cambridge University Press, 2012.

Outler, Albert. "The Wesleyan Quadrilateral in John Wesley." *Wesleyan Theological Journal* 21.1 (Spring 1985): 7-18.

Page, Ruth. "Panentheism and Pansyntheism: God in Relation." *In Whom We Live and Move and Have Our Being: Panentheistic Reflections on God's Presence in a Scientific World*, 222–32. Edited by Philip Clayton and Arthur Peacocke. Grand Rapids: Eerdmans, 2004.

Pals Daniel L. *Eight Theories of Religion*. 2nd Edition. New York: Oxford University Press, 2006.

Pannenberg, Wolfhart. *Systematic Theology*. Translated by Geoffrey W. Bromiley. Grand Rapids: Eerdmans, 1991.

Pascal, Blaise. *Pensées*. Translated by A. J. Krailsheimer. Harmondsworth: Penguin, 1966.

Peacocke, Arthur. "Articulating God's Presence in and to the World Unveiled by the Sciences." *In Whom We Live and Move and Have Our Being*, 137–154. Edited by Philip Clayton and Arthur Peacocke. Grand Rapids: Eerdmans, 2004.

_______. *Creation and the World of Science*, Oxford: Clarendon Press, 1979.

_______. "Emergence, Mind, and Divine Action: The Hierarchy of the Sciences in Relation to the Human Mind-Brain-Body," 257–78. *The Re-Emergence of Emergence: The Emergentist Hypothesis from Science to Religion*. Edited by Philip Clayton and Paul Davies. Oxford: Oxford University Press, 2006.

_______. *Paths from Science towards God: The End of All Our Exploring*. New York: Oneworld, 2001.

Pennock, Robert T., ed. *Intelligent Design Creationism and Its Critics: Philosophical, Theological, and Scientific Perspectives*. Cambridge: MIT Press, 2001.

_______. *Tower of Babel: The Evidence against the New Creationism*. Cambridge: MIT Press, 1999.

Peters, Ted, ed. *Science and Theology*. Boulder: Westview, 1999.

Pickstock, Catherine. *After Writing: On the Liturgical Consummation of Philosophy*. Oxford: Blackwell Publishers, 1998.

Plantinga, Alvin. "Creation and Evolution: A Modest Proposal." *Intelligent Design Creationism and Its Critics: Philosophical, Theological, and Scientific Perspectives*, 779-92. Edited by Robert T. Pennock. Cambridge: MIT Press, 2001.

_______. "Methodological Naturalism?" *Intelligent Design Creationism and Its Critics: Philosophical, Theological, and Scientific Perspectives,* 339-62. Edited by Robert T. Pennock. Cambridge: MIT Press, 2001.

_______. "Methodological Naturalism?" *Origins and Design* 18.1 (1997): 1-17.

_______. "Methodological Naturalism? Part 2." *Origins and Design* 18.2 (1997): 1-19.

_______. *Where the Conflict Really Lies: Science, Religion, & Naturalism*. Oxford: Oxford University Press, 2011.

Plantinga, Alvin and Nicholas Wolterstorff, eds. *Faith and Rationality: Reason and Belief in God*. Notre Dame: University of Notre Dame Press, 1983.

Poincaré, Henri. *Science and Method*. New York: Dover, 2003.

Polkinghorne, John. *Quantum Physics and Theology: An Unexpected Kinship*. New Haven: Yale University Press, 2007.

_______. *Science and Religion in Quest of Truth*. New Haven: Yale University, 2011.

_______. "The Demise of Democritus." *The Trinity and the Entangled World: Relationality in Physical Science and Theology*, 1–14. Edited by John Polkinghorne. Grand Rapids: Eerdmans, 2010.

Pope, Stephen J. *Human Evolution and Christian Ethics.* Cambridge, UK: Cambridge University Press, 2007.

Porter, Steven L. "Rethinking the Logic of Penal Substitution." *Philosophy of Religion: A Reader and Guide*, 596–608. Edited by William Lane Craig. New Brunswick: Rutgers University Press, 2002.

Price, Lucien. *Dialogues of Alfred North Whitehead.* New York: Mentor Books, 1956.

Quinn, Warren. "The Right to Threaten and the Right to Punish." *Philosophy & Public Affairs* 14.4 (1985): 327–73.

Rahner, Karl. *Foundations of Christian Faith: An Introduction to the Idea of Christianity*. Translated by William V. Dych. New York: Crossroads, 1978.

Ramachandran, Vilayanur S. *The Tell-Tale Brain: A Neuroscientist's Quest for What Makes Us Human*. New York: W. W. Norton and Sons, 2011.

Rawls, John. "Two Concepts of Rules." *The Philosophical Review* 64.1 (January 1955): 3-32.

Riis, Søren. "Towards the Origin of Modern Technology: Reconfiguring Martin Heidegger's Thinking." *Continent-al Philosophy Review* 44.1 (2011): 103-117.

Rozin, Paul, Laura Lowery, and Jonathan Haidt. "The CAD Triad Hypothesis: A Mapping Between Three Moral Emotions (Contempt, Anger, Disgust) and Three Moral Codes (Community, Autonomy, Divinity)." *Journal of Personality and Social Psychology* 76.4 (1999): 574–86.

Ruse, Michael. "Methodological Naturalism Under Attack." *South African Journal of Philosophy* 24.1 (2005): 44-60.

Russell, Robert J. *Time in Eternity: Pannenberg, Physics, and Eschatology in Creative Mutual Interaction.* Notre Dame: University of Notre Dame, 2012.

_______. "Bridging Science and Religion: Why It Must Be Done." *Theology and Science* 1.1 (2003): 1–3.

Russell, Robert J., Nancey Murphy, Theo C. Mayering and Michael A. Arbib, eds. *Neuroscience and the Person: Scientific Perspectives on Divine Action*. Vatican City: Vatican Observatory along with the Center for Theology and the Natural Sciences, Berkeley, 1999.

Russell, Robert. *John Paul II on Science and Religion: Reflections on the New View from Rome.* Vatican City: Vatican Observatory, 1990.

Sacks, Oliver. *Hallucinations*. New York: Knopf, 2012.

Shafer-Landau, Russ. *Moral Realism: A Defence*. Oxford: Oxford University Press, 2003.

Sharpe, Eric. J. *Comparative Religion: A History*. LaSalle: Open Court, 1986.

Shermer, Michael. "Proof of Hallucination." *Scientific American* 308.4 (April 2013): http://www.michaelshermer.com/2013/04/proof-of-hallucination/.

Siegle, Daniel J. *The Mindful Brain: Reflection and Attunement in the Cultivation of Well-Being*. New York: W. W. Norton, 2007.

Sklar, Jay. *Sin, Impurity, Sacrifice, Atonement: The Priestly Conceptions*. Sheffield: Sheffield: Phoenix Press, 2005.

Smith, John E., Harry S. Stout, and Kenneth P. Minkem, eds. *A Jonathan Edwards Reader*. New Haven: Yale University Press, 2003.

Sober, Elliott. "Why Methodological Naturalism?" *Biological Evolution: Facts and Theories: A Critical Appraisal 150 Years after The Origin of Species*, 359-78. Edited by Gennaro Aulette, Marc LeClerc, and Rafael Martinez. Rome: Gregorian Biblical Press, 2011.

Sokolowski, Robert. *The God of Faith and Reason: Foundations of Christian Theology*. Washington, DC: Catholic University of America Press, 1995.

Souza, David A., ed. *Mind, Brain, and Education: Neuroscience Implications for the Classroom*. Bloomington: Leading Edge, Solution Tree, 2010.

Spaulding, Henry, II. "Good Conscience or Good Confidence: A Postmodern Re-thinking of Ethical Reflection in the Wesleyan/Holiness Tradition." *Wesleyan Theological Journal* 35.1 (Spring 2000): 41-66.

Srivastava, Joydeep, and Francine Espinoza. "Coupling and Decoupling of Unfairness and Anger in Ultimatum Bargaining." *Journal of Behavioral Decision Making* 22 (2009): 475–89.

Stoeger, William R. "Contemporary Physics and the Ontological Status of the Laws of Nature." *Quantum Cosmology and the Laws of Nature: Scientific Perspectives on Divine Action*, 209–34. Edited by Robert J. Russell, Nancey C. Murphy and Chris J. Isham. Vatican City: Vatican Observatory Publications, 1993.

Stolnitz, Jerome. "Beauty." *The Encyclopedia of Philosophy*. New York: MacMillan and Free Press, 1967.

Stott, John R. W. *The Cross of Christ*. Madison: Inter-Varsity Press, 1986.

Sznycer, Daniel, Eric Schniter, John Tooby, and Leda Cosmides. "Regulatory Adaptations for Delivering Information: The Case of Confession." *Evolution and Human Behavior* 36.1 (2015): 44–51.

Taves, Anne. *Religious Experience Reconsidered: A Building-Block Approach to the Study of Religion and Other Special Things*. Princeton: Princeton University Press, 2010.

Taylor, Mark C. "Introduction." *Critical Terms for Religious Studies*, 1-19. Edited by Mark C. Taylor. Chicago: University of Chicago Press, 1998.

Thomas, Owen C. "Problems in Panentheism." *The Oxford Handbook of Religion and Science*, 652–64. Edited by Philip Clayton. Oxford: Oxford University Press, 2006.

Tokuhama-Espinosa, Tracey. *Mind, Brain, and Education Science: A Comprehensive Guide to the New Brain-Based Teaching*. New York: Teachers College Press, 2010.

Varghese, Roy Abraham, ed. *The Intellectuals Speak Out About God*. Chicago: Regnery Gateway, 1984.

von Balthasar, Hans Urs. *The Glory of the Lord: A Theological Aesthetics*. 7 vols. Edinburgh: T & T Clark, 1982-91.

Wainwright, William. "Concepts of God." *The Standford Encyclopedia of Philosophy*. Spring 2013 Edition.

Walton, John. *The Lost World of Genesis One: Ancient Cosmology and the Origins Debate*. Downers Grove: IVP, 2009.

Ward, Keith. *The Big Questions in Science and Religion*. West Conshohocken: Templeton, 2008.

Welker, Michael. *Creation and Reality*. Translated by John F. Hoffmeyer. Minneapolis: Fortress, 1999.

Wesley, John. *A Survey of the Wisdom of God in Creation, or A Compendium of Natural Philosophy*. 2 vols. 1st ed. Bristol: William Pine, 1763.

________. *A Survey of the Wisdom of God in Creation, or A Compendium of Natural Philosophy*. 5 vols. 1st Edition. London: J. Paramore, 1784.

________. *The Works of John Wesley*. Edited by Thomas Jackson. London: Wesleyan Methodist Book Room, 1872.

_______. *The Works of the Rev. John Wesley: Forty-Two Sermons on Various Subjects*, Vol. 5. New York: J & J Harper, 1826.

_______. *John Wesley*. Edited by Albert Cook Outler. New York: Oxford University Press, 1964.

Whitehead, Alfred North. *Adventures of Ideas*. New York: Free Press, 1933.

_______. *Aims of Education and Other Essays*. New York: Free Press, 1967.

_______. *An Enquiry Concerning the Principles of Natural Knowledge*. Cambridge: Cambridge University Press, 1925.

_______. *Modes of Thought*. New York: Free Press, 1968.

_______. *Process and Reality: An Essay in Cosmology*. New York: Free Press, 1978.

_______. *Process and Reality*. Corrected ed. Edited by David Ray Griffin and Donald W. Sherburne. New York: Free Press, 1978.

_______. *Religion in the Making*. New York: Meridian Books, 1960.

_______. *Science and the Modern World*. New York: Free Press, 1967.

_______. *The Concept of Nature*. Cambridge: University Press, 1971.

Wiegman, Isaac. "The Evolution of Retribution: Intuitions Undermined." *Pacific Philosophical Quarterly* 92.2 (2015): 1-26.

_______. "Divine Retribution in Evolutionary Perspective." *In Spirit and Truth: Philosophical Reflection on Liturgy and Worship*, 181-202. Edited by Wm. Curtis Holtzen and Matthew Nelson Hill. Claremont: Claremont Press, 2016.

Wildman, Wesley J. "Mark Johnston's Naturalistic Account of God and Nature, Life and Death." *American Journal of Theology and Philosophy* 32.2 (2011): 180-87.

Willis, Judy. *Research Based Strategies to Ignite Student Learning: Insights from a Neurologist and Classroom Teacher.* Alexandria, VA: Association for Supervision & Curriculum Development, 2006.

Wilson, David B. *Seeking Nature's Logic: Natural Philosophy in the Scottish Enlightenment.* University Park: Pennsylvania State University Press, 2009.

Wittgenstein, Ludwig. *Philosophical Investigations.* Chichester, UK: Basil Blackwell, 1953.

Wolterstorff, Nicholas. *Works and Worlds of Art.* New York: Clarendon Press, 1980.

Worthing, Mark William. *God, Creation, and Contemporary Physics.* Minneapolis: Fortress Press, 1996.

## *Index of Authors*

## *Index of Subjects*

Made in the USA
Monee, IL
04 November 2026

67382989R00177